TSUNAMI BEACHES

A tale of love and loss

GUSTAV WIK

ISBN: 1461017947
ISBN-13: 9781461017943

"The only silence we know is the silence when noise stops, the silence when thought stops – but that is not silence. Silence is something entirely different, like beauty, like love....

It cannot be described. What can be described is the known, and the freedom from the known can come into being only when there is a dying every day to the known, to the hurts, the flatteries, to all the images you have made, to all your experiences – dying every day so that the brain cells themselves become fresh, young, innocent....

That silence which is not the silence of the ending of noise is only a small beginning. It is like going through a small hole to an enormous, wide, expansive ocean, to an immeasurable, timeless state."

In: Freedom from the Known, J. Krishnamurti, Victor Gollancz Ltd, London, 1972

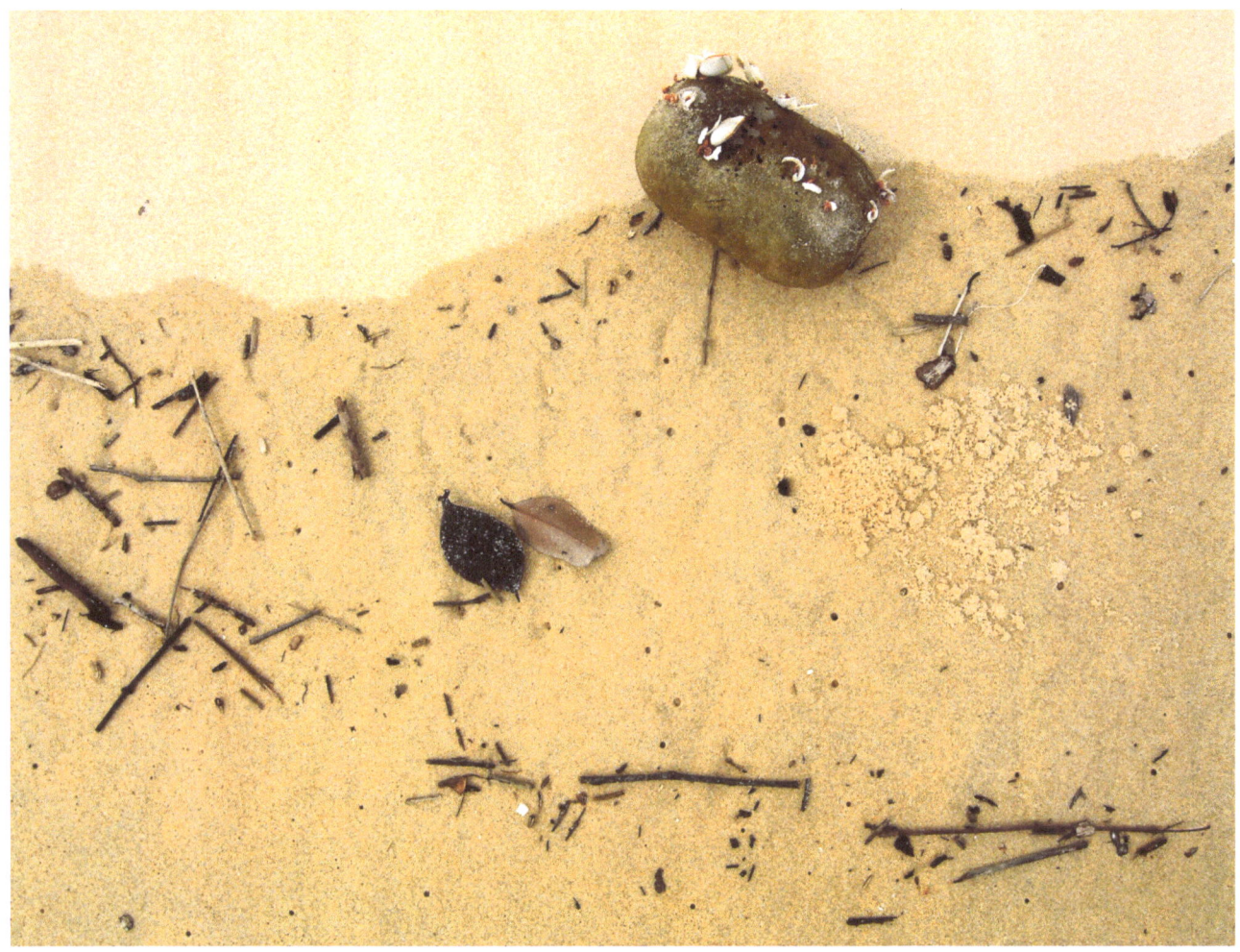

In totality there is no right and wrong

The writing hereunder the pictures is about mental injuries. It is about scars in your heart coming from sudden shocks of terror or from losing your dear ones. It is about losing sanity, identity, and reasons to live. It is about those who lived along the Indian Ocean coasts where the Christmas 2004 Tsunami destroyed so much and killed so many. These harmful acts of God or man are part of life. They lure there in our future, or maybe also in our past, to hit us under every likely or unlikely circumstance. They appear out of nothing or results from provocations. They come out of the ordinary or the bizarre, the calm or the chaotic, peace or war.

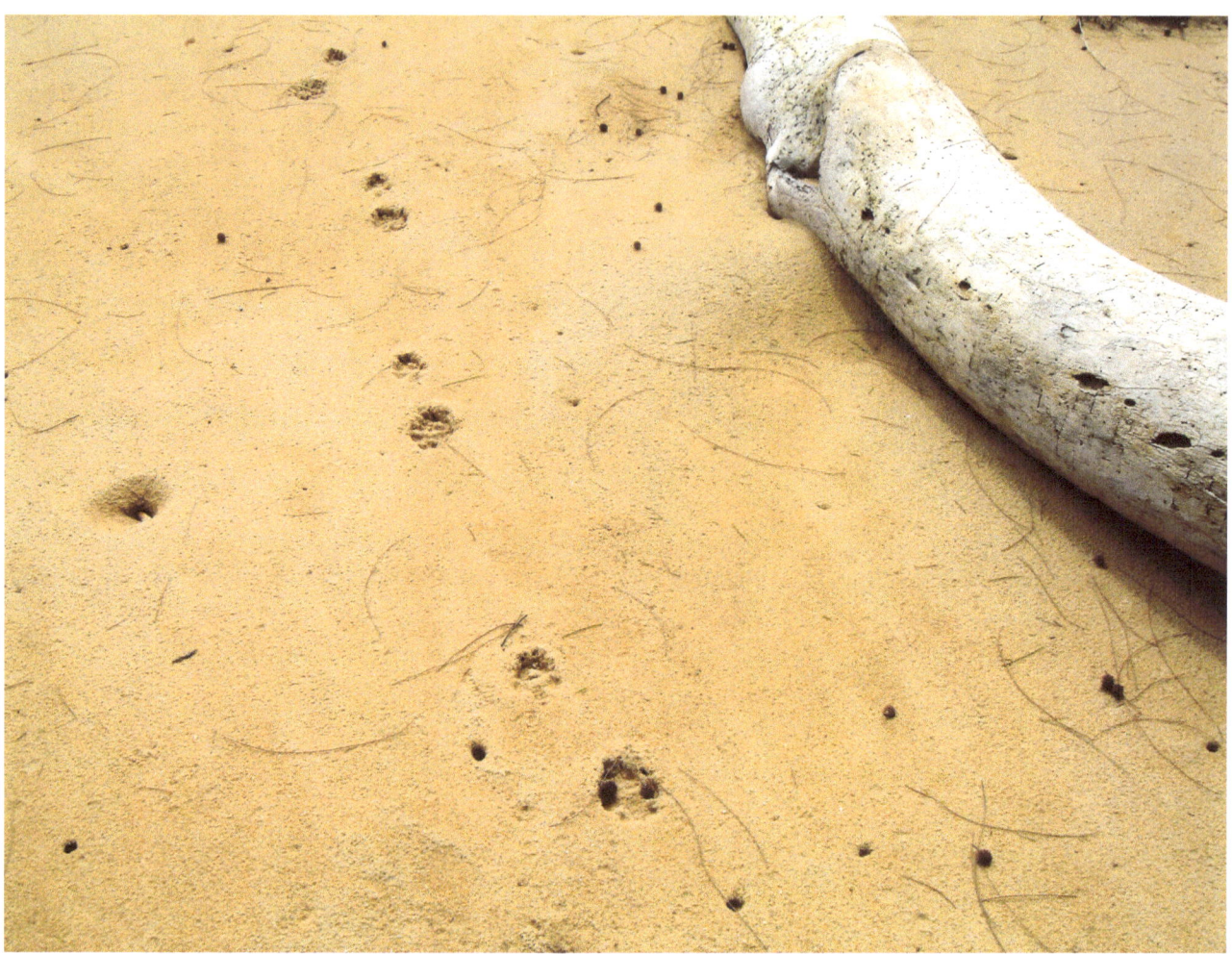

There is no escape from actual facts

Chiraphorn took me to hard-to-believe scenes of that God's unforeseen act some six month after the tsunami. Without any warning, God let a monstrous wave strike innocent humans and other living creatures, peacefully dwelling along paradise-like beaches.

You often do not even notice a tsunami in the open sea. It instead releases its deadly forces when it hits the shore. Japanese fishermen gave us the word. They did not notice much at sea, but coming back home they saw their village annihilated. The English translation for tsunami would be "harbour-wave."

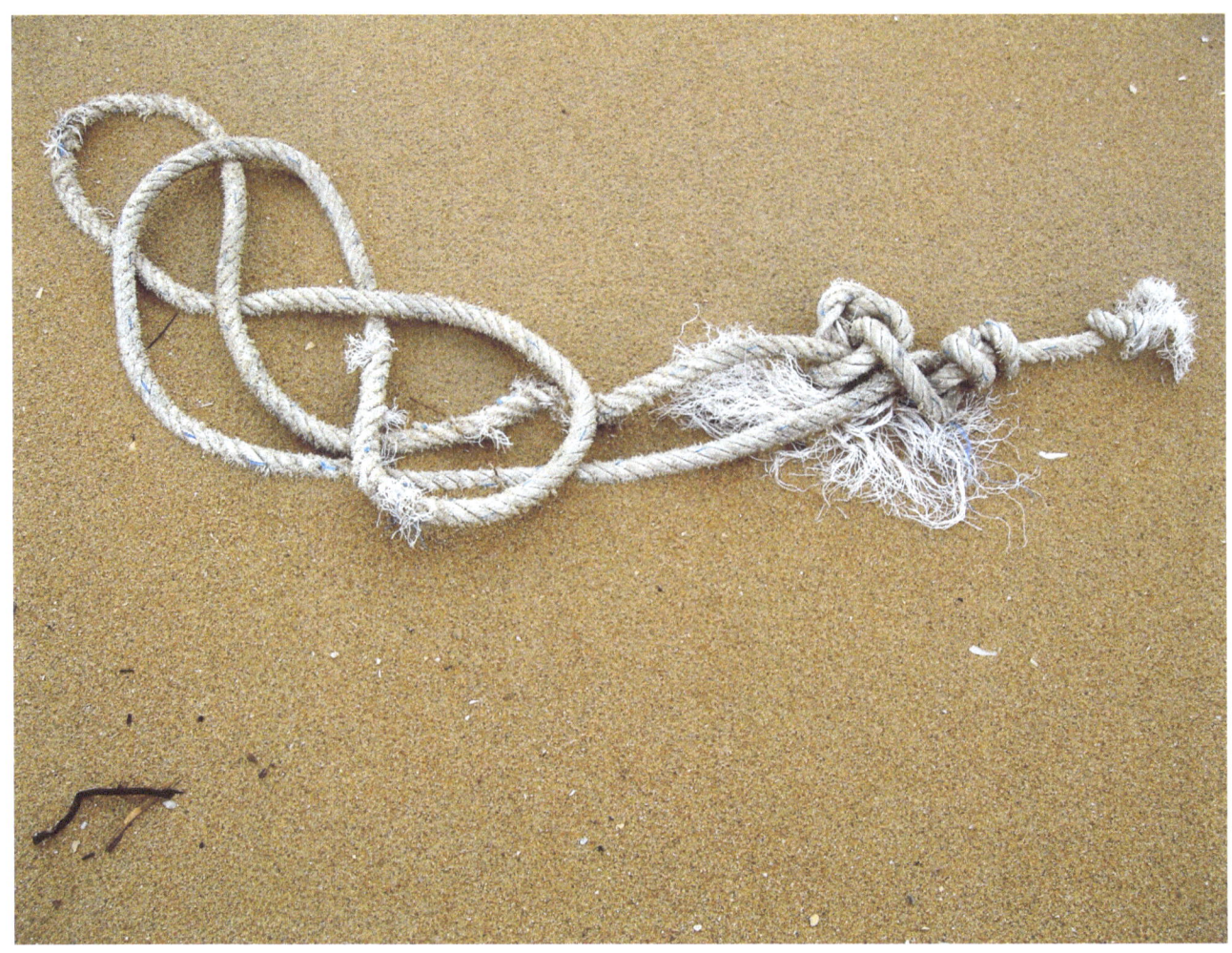

Why do we always compare what we are with what we should be?

At 07:58:53 local time December 26, 2004 seismic activity was registered off the coast of Sumatra in Indonesia. Subsequent movements of sub-sea landmasses set off the tsunami, and not more than minutes later, the monster-wave killed about 300,000 humans and many millions of other living creatures. Close to the earthquake; in Indonesia, Thailand, and Malaysia the unhindered wave slay most of its victims. The annihilation of life continued much farther though; in Bangladesh, India, Sri Lanka, the Maldives, and even in Eastern Africa.

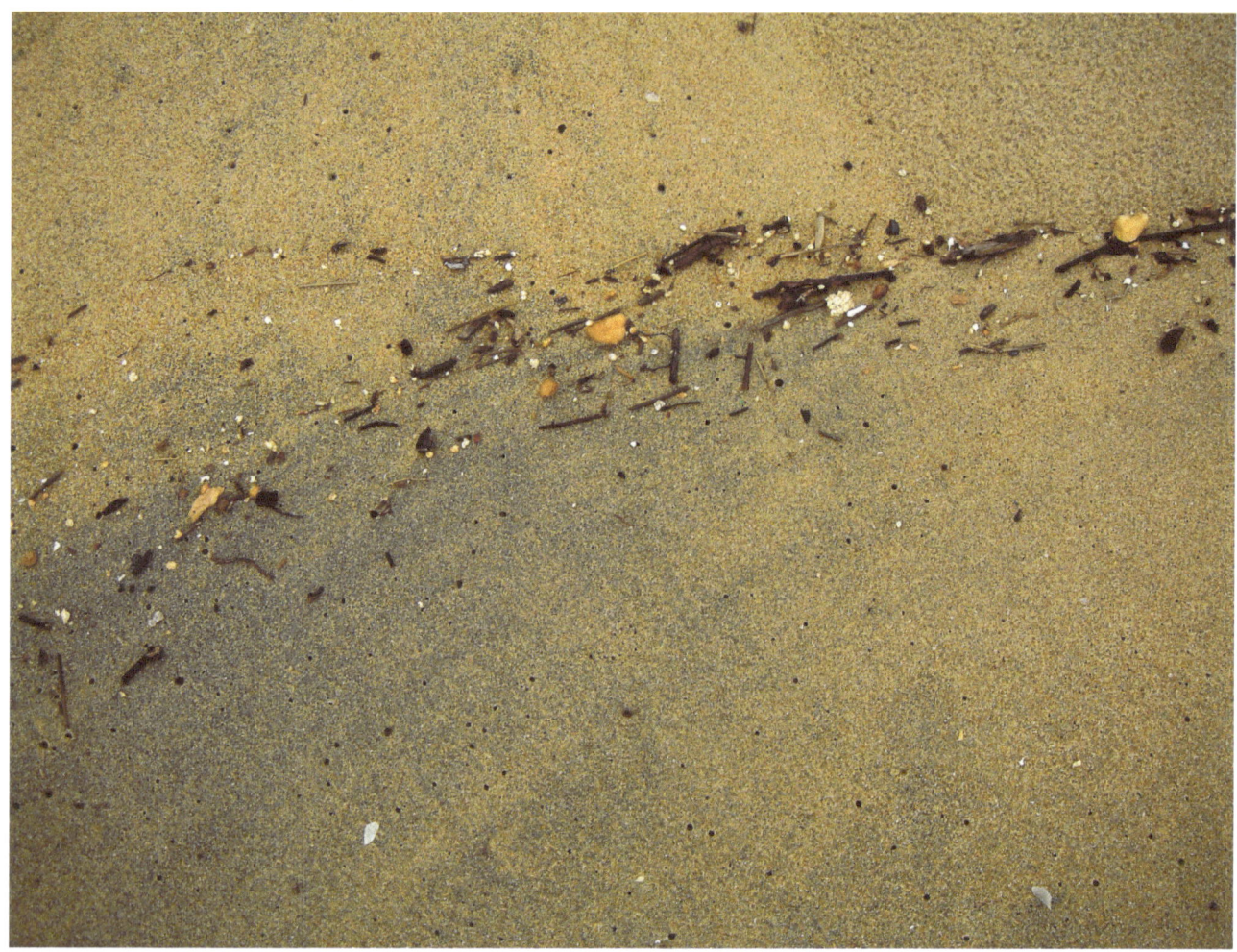

Maybe there is a line in the sand

Why should we talk about numbers? So many too young and sound people have died in this world, from accidents, disasters, epidemics, wars, murder, suicide and all other possible causes. Do you not believe that far too many healthy humans will continue to die like that?

Let us go back just some years. What about the Second World War (1939 to 1945)? More than 60 million people died here. If we to that add the Second Sino-Japanese War (1931–1945), it comes to 80 million. Do we still call those times "modern and civilized," or have we forgotten about them? They say Neil Armstrong, one man only, walked on the moon in 1964. Is that true?

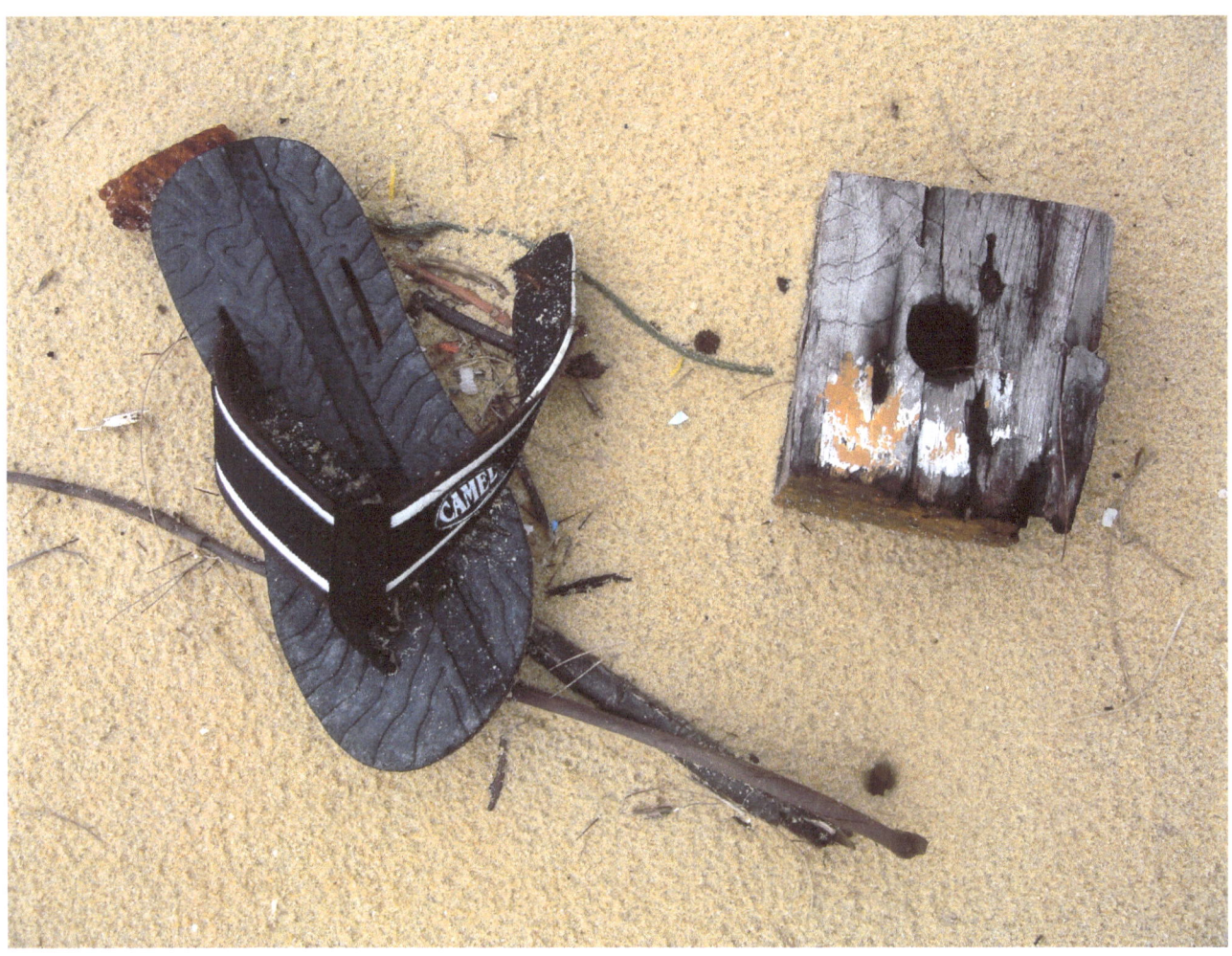

Resonance in black

Maybe times have not changed so much. We still have famines, epidemics, cyclones, floods, and a seemingly never-ending "war against terrorism." Who will be our next foe?

This telling is anyhow not about numbers, but about some few individuals only. Ultimately it is about the survival of one little girl and the hope she gave others. If you ask, "What is this one person's survival against all those other's who died?" I will answer, "Numbers are nonexistent. It is about real against relative. Life is real, whereas numbers are artificial constructions."

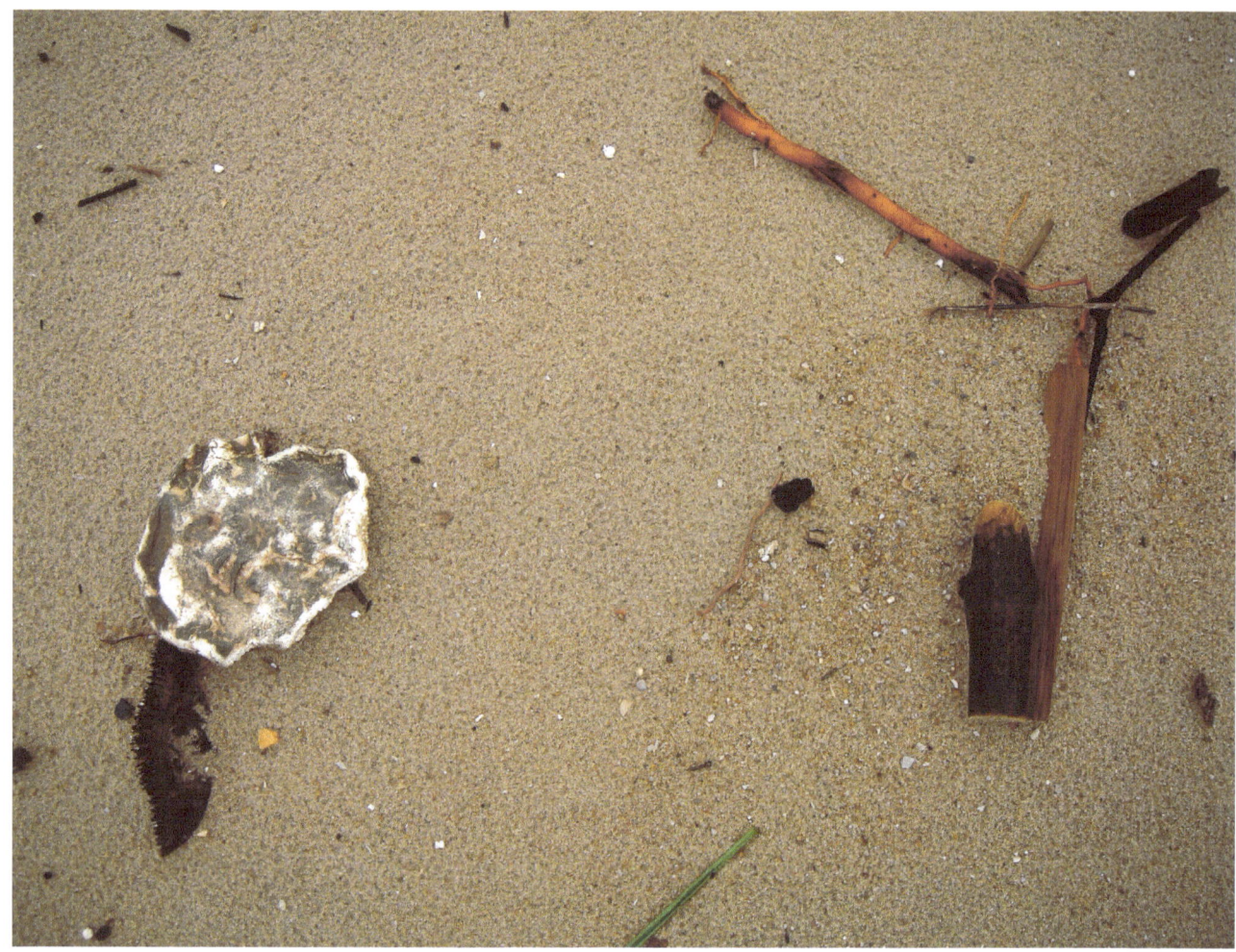

Group patterns will soon change – No harm

The Indian Ocean Tsunami became real to Europeans and Americans because their family and friends died here. Love and loss is real whereas a given number of deaths have little impact.

For us in the West, the village of Khao Lak became as a symbol of the disaster. Khao Lak appeared in my morning paper every day, month after month. Everyone talked about the tragedy. At times sorrow turned into anger. The press and public meant that our officials had not taken initial information serious, had not acted promptly enough. Emotions were aroused. If I recall it correctly, ministers and officials had to resign, even in far-away Scandinavia.

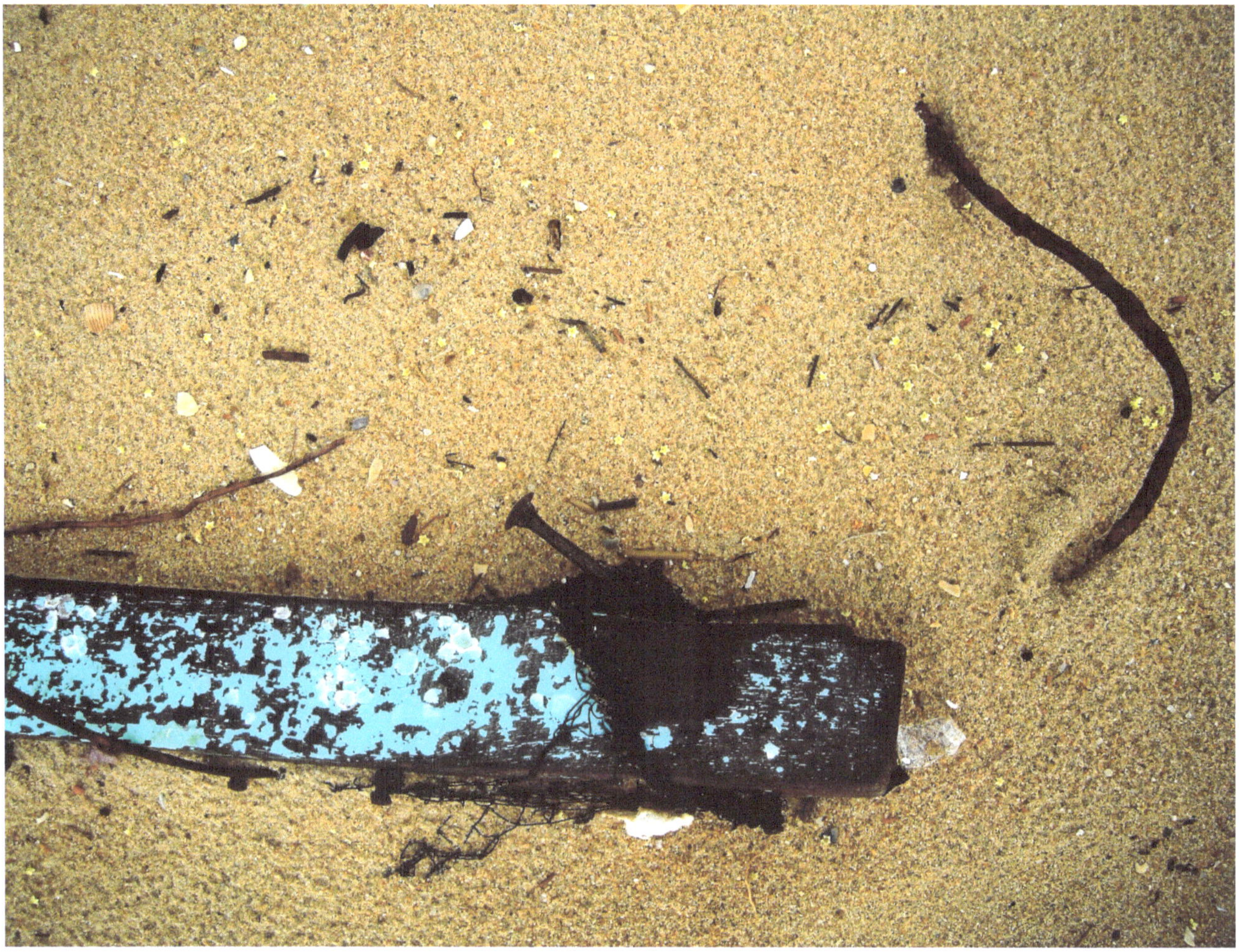

This is one of the colors of truth and love

When we came to Khao Lak there were no tourists there of course, and most of all others who had been there to help and assist were gone back home too. Rebuilding had begun but the scenes of destruction were still there, expressing nature's ultimate power.

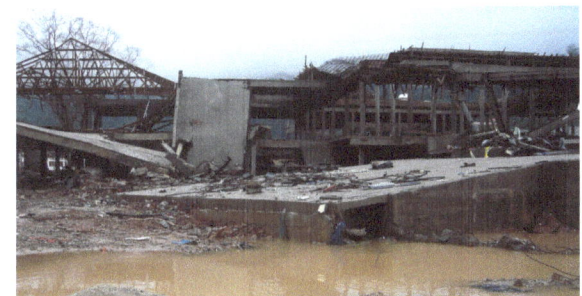

Nature is so powerful

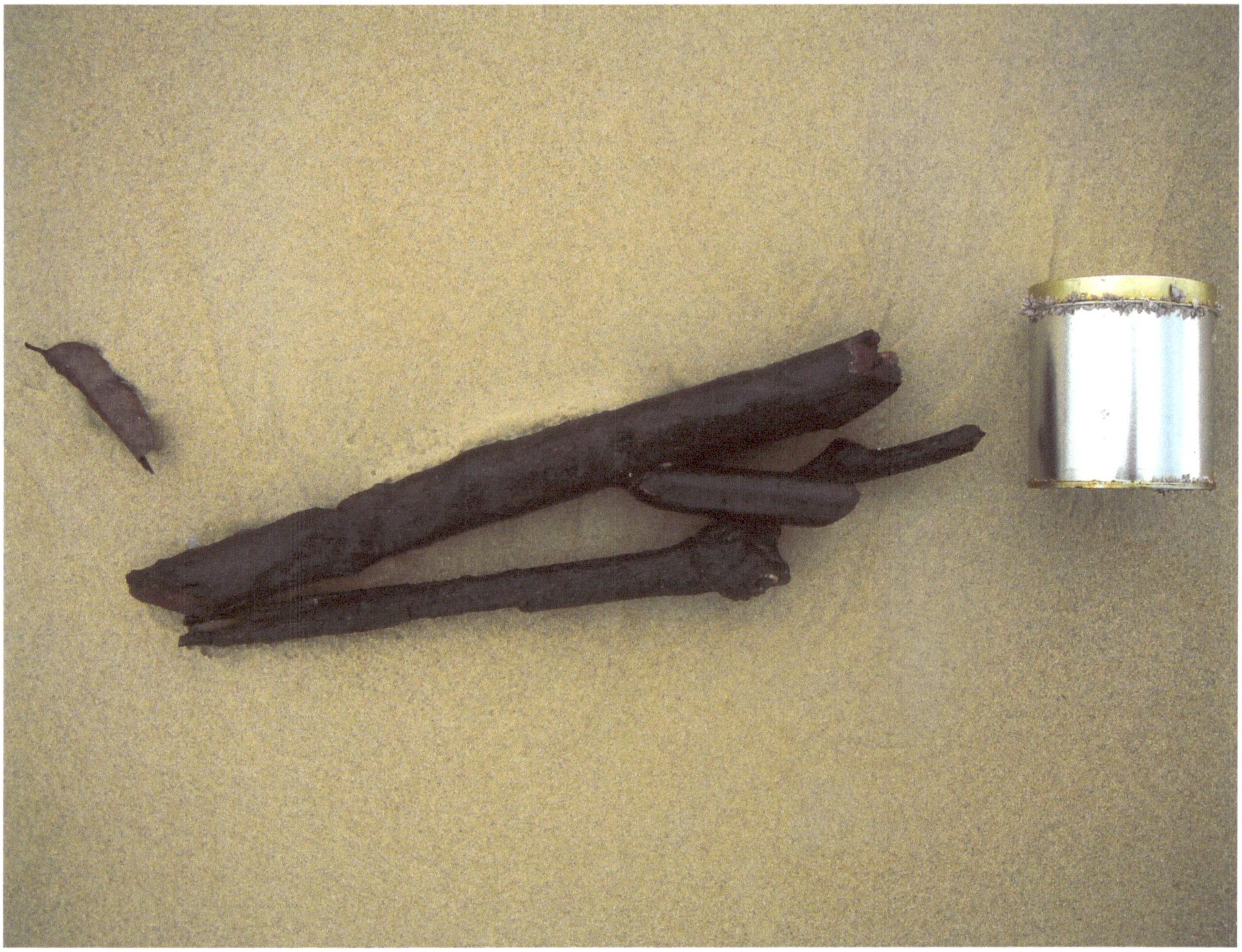

![photograph of charred debris and a metal can on sand]

Do not try to resolve fear

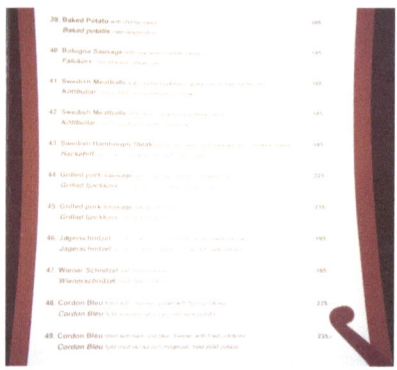

No ordinary man's menu

After walking along the beaches for quite some time, taking these pictures, I got hungry and strongly wanted lunch. We went to one of the very few Khao Lak restaurants still there, saved by a location somewhat higher above the sea. The staff who spoke English was now at more affluent places, like Pattaya or Koh Samui. The menu was still the same however, not expressed in Thai, but only in English and Swedish.

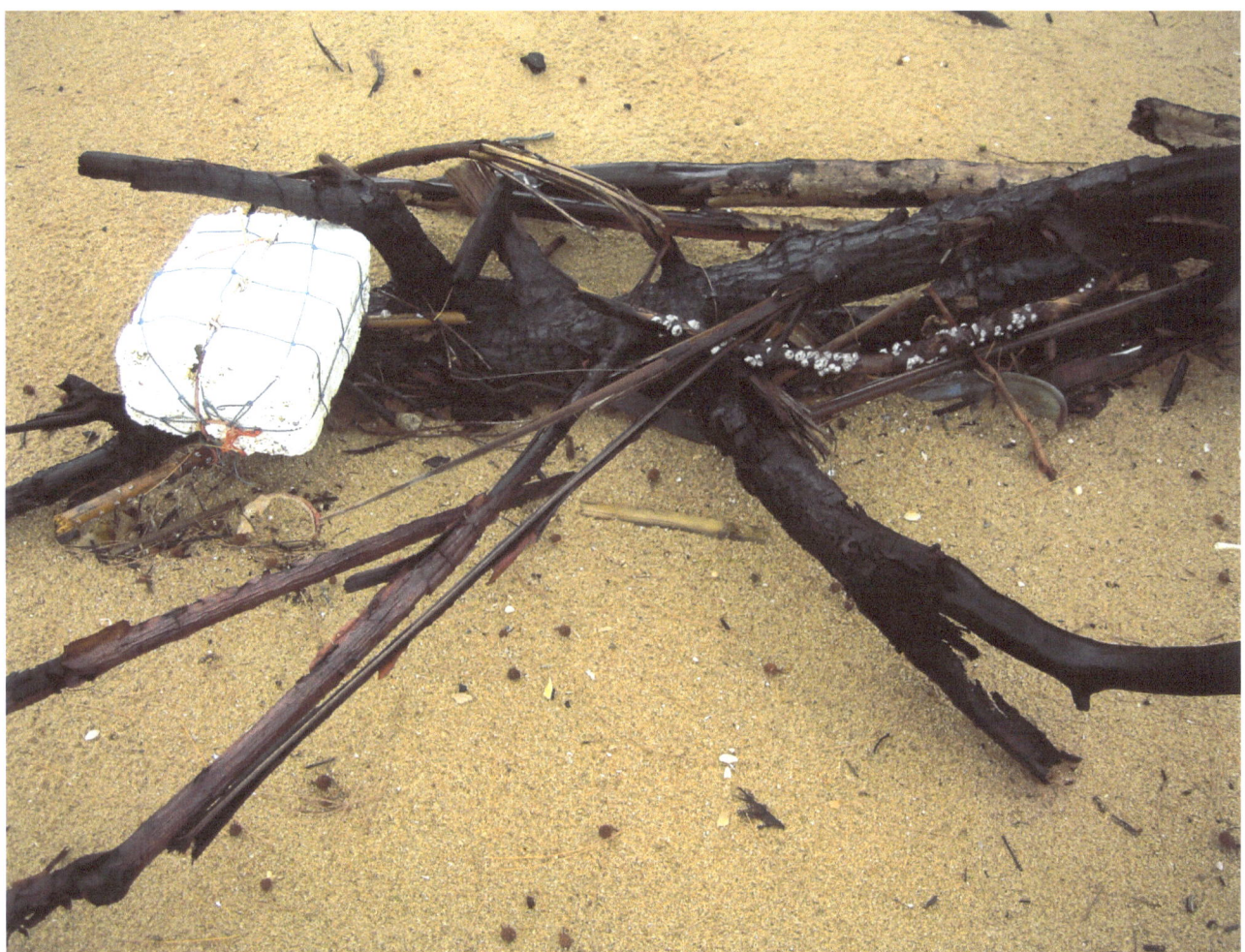

Floating and semi-floating materials

I am of Scandinavian and North European origin. A **Filet Mignon steak** with Roquefort Sauce or Swedish Meatballs with mashed potato would have been superb for lunch. Chiraphorn had never tasted such food however and asked for Thai mango salad, which they did not have. I wanted us to stay at the forlorn restaurant, but had to accept Chiraphorn's stronger will. My spirits dampened and very hungry I reasoned, *No fresh food in this forlorn place anyhow. Those meatballs were certainly deep-frozen long before the monster appeared.*

This mid-July rainy day another abandoned and frozen scene appeared outside the restaurant.

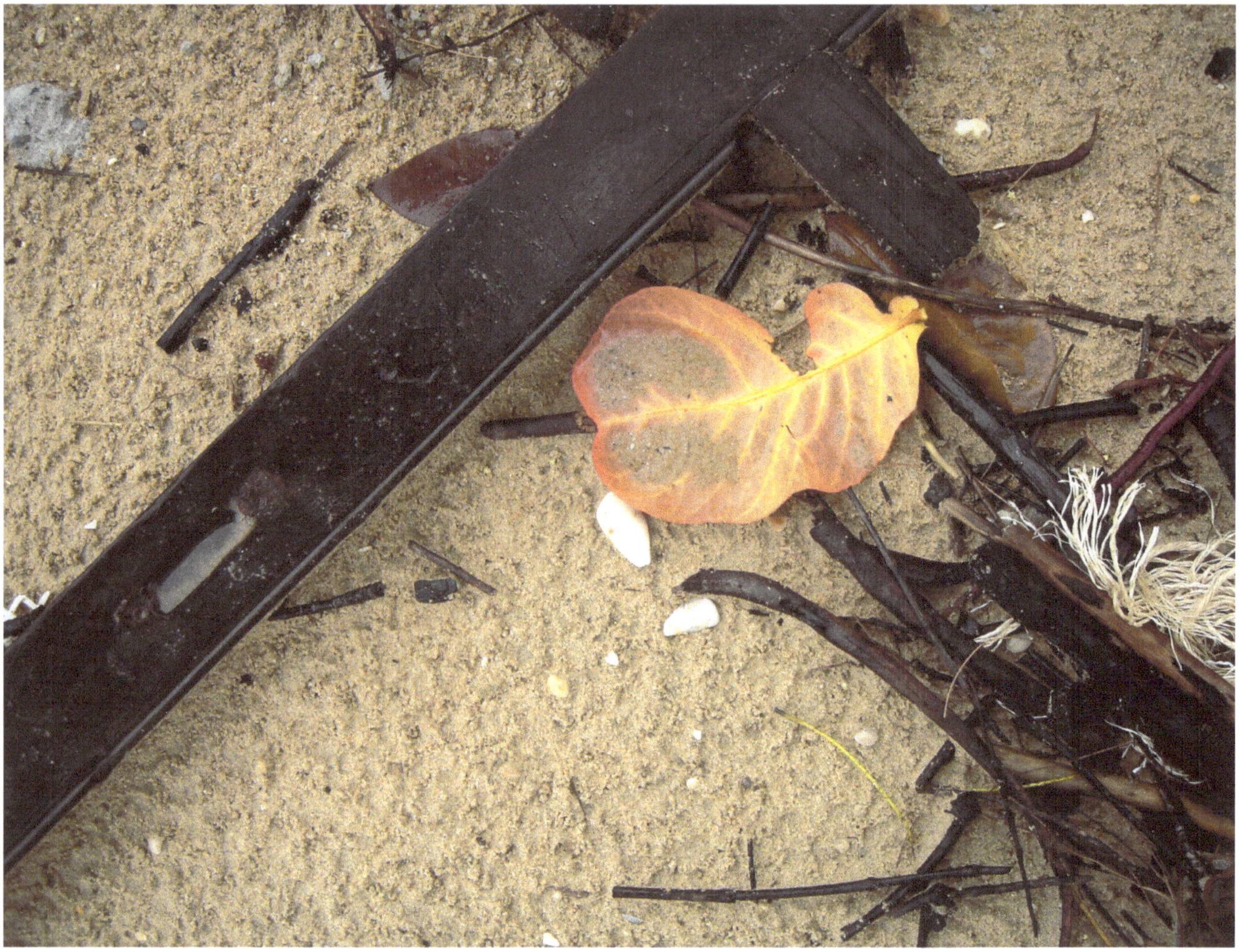

![photograph of a leaf and debris on sand]

It is not a cross but a leaf

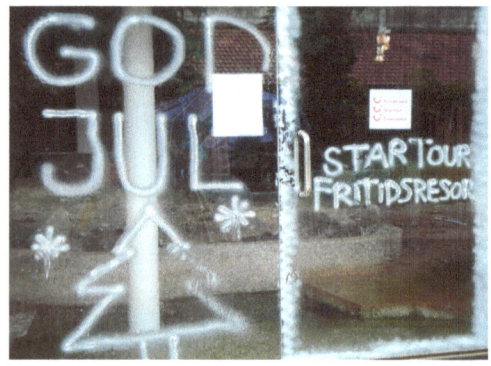

It says Merry Christmas

In my own exotic language, the display window of a closed-down travel agency still proclaimed a "Merry Christmas" in snowy white. Chiraphorn could not read the Scandinavian text and its greeting, but turned to me, "If you want to grasp the sentiments of the people here, those who lost everything, from whom the tsunami took the lot, you shouldn't stay in Khao Lak, but let me take you to a village where nothing was left.

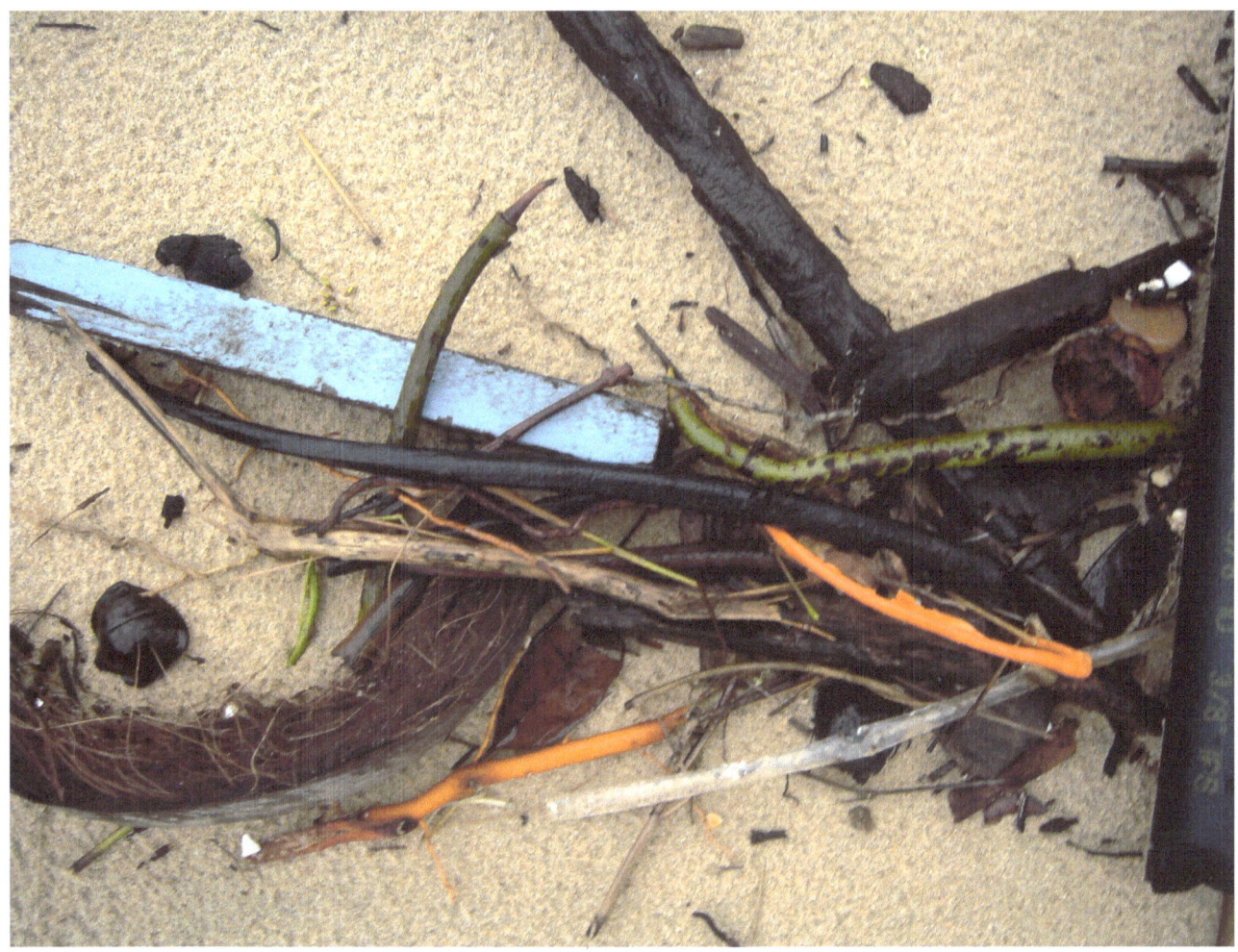

Woods and plastic

"Understand too, white man, that this complete loss is true for so many thousands other settlings around the Indian Ocean. Only you did not read about those places in your part of the world. Few were interested in those locations as all those other people's losses were unrelated to your kin."

At about noontime we drove further north up the coast. About an hour's ride later the road became narrower. We had to drive quite slow for another twenty minutes, before we turned left onto a broad but provisional gravel road, made for heavy vehicles. After the left turn, we headed west towards the sea and soon arrived to the actual village Chiraphorn wanted me to see.

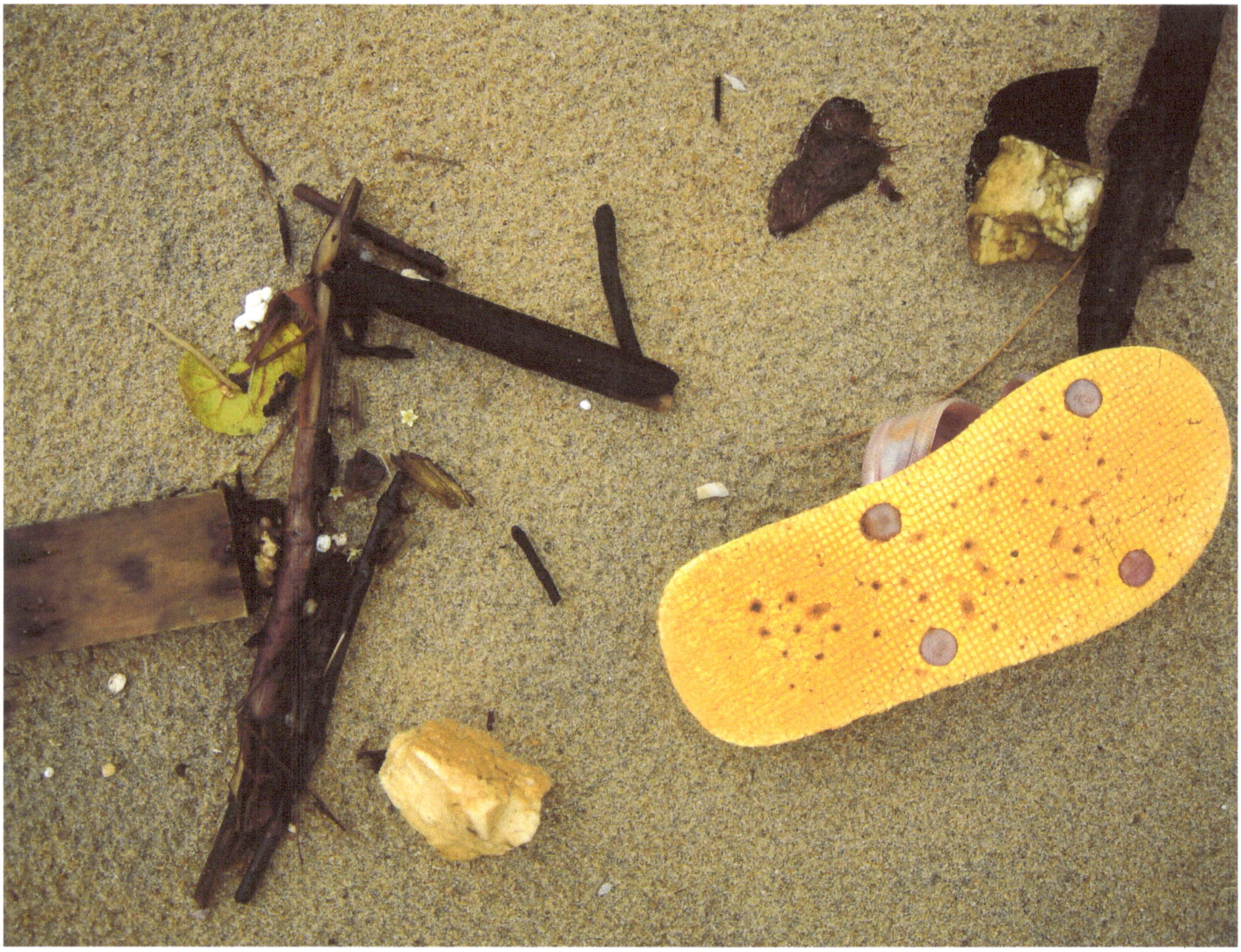

No more talking

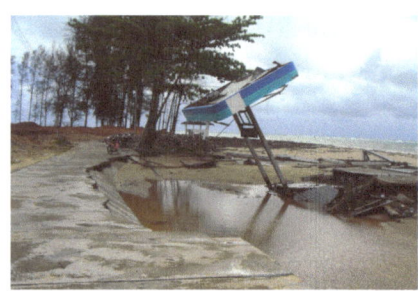

A former gas station

Maybe it once was an idyllic village. Now it looked like a military camp however, with young soldiers building houses everywhere around. We parked the car by the remnants of a gas station. Chiraphorn looked at me. "Around here used to be some very beautiful homes." As in an inner vision, a second vicious wave approached from the sea. A flash of panic struck me.

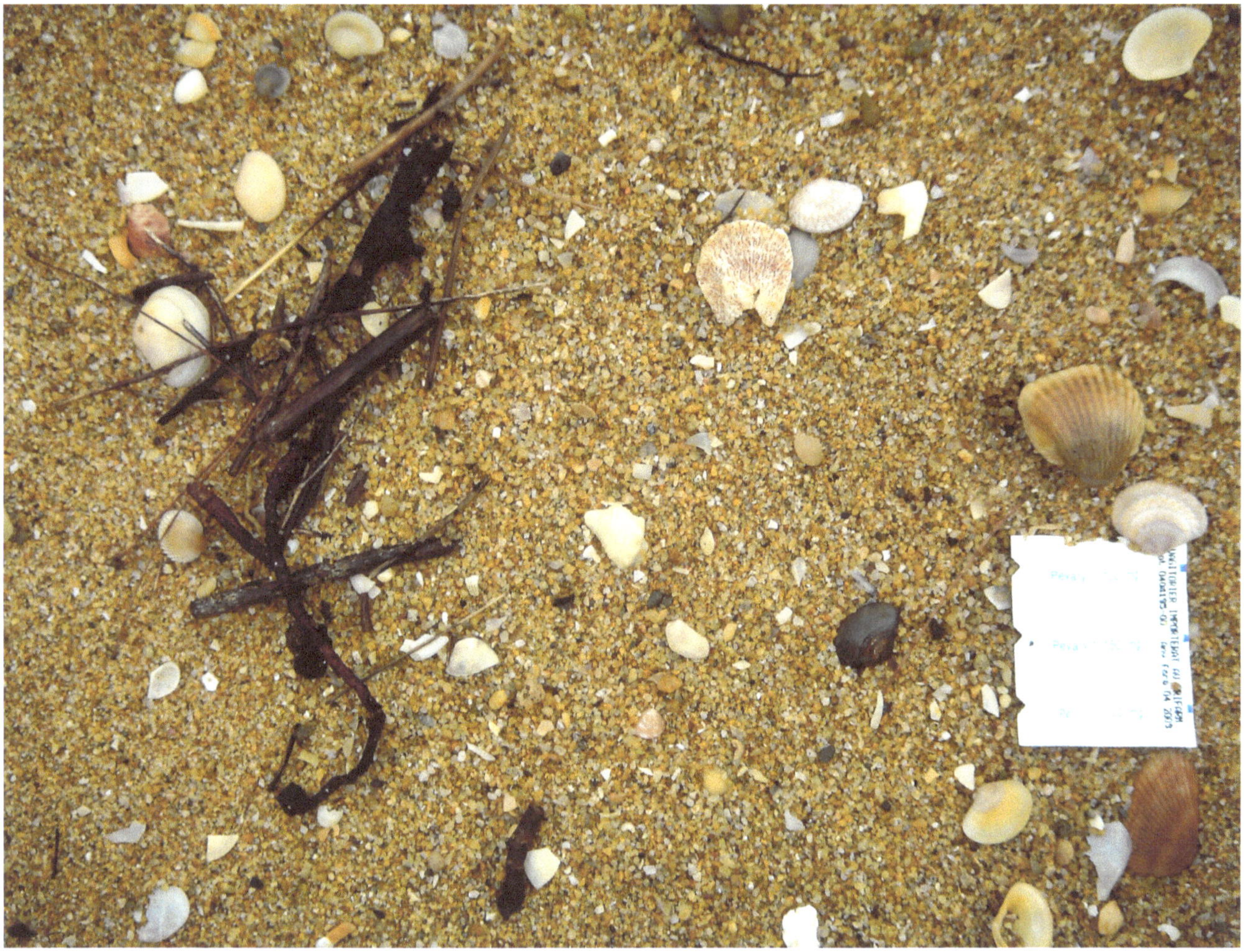

Shells and medications

We walked around a little more in that rebuilding activity. I took a picture of a Buddha icon at the harbour, and pondered if the icon could have been there during the act of destruction. I realized though that people must have brought it there later, to give consolation and hope. The Buddha icon stood at harbour entrance. If you turn around to see what the icon sees,

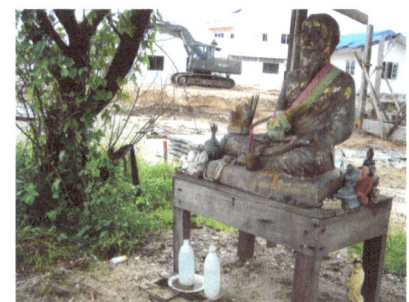

There to give consolation

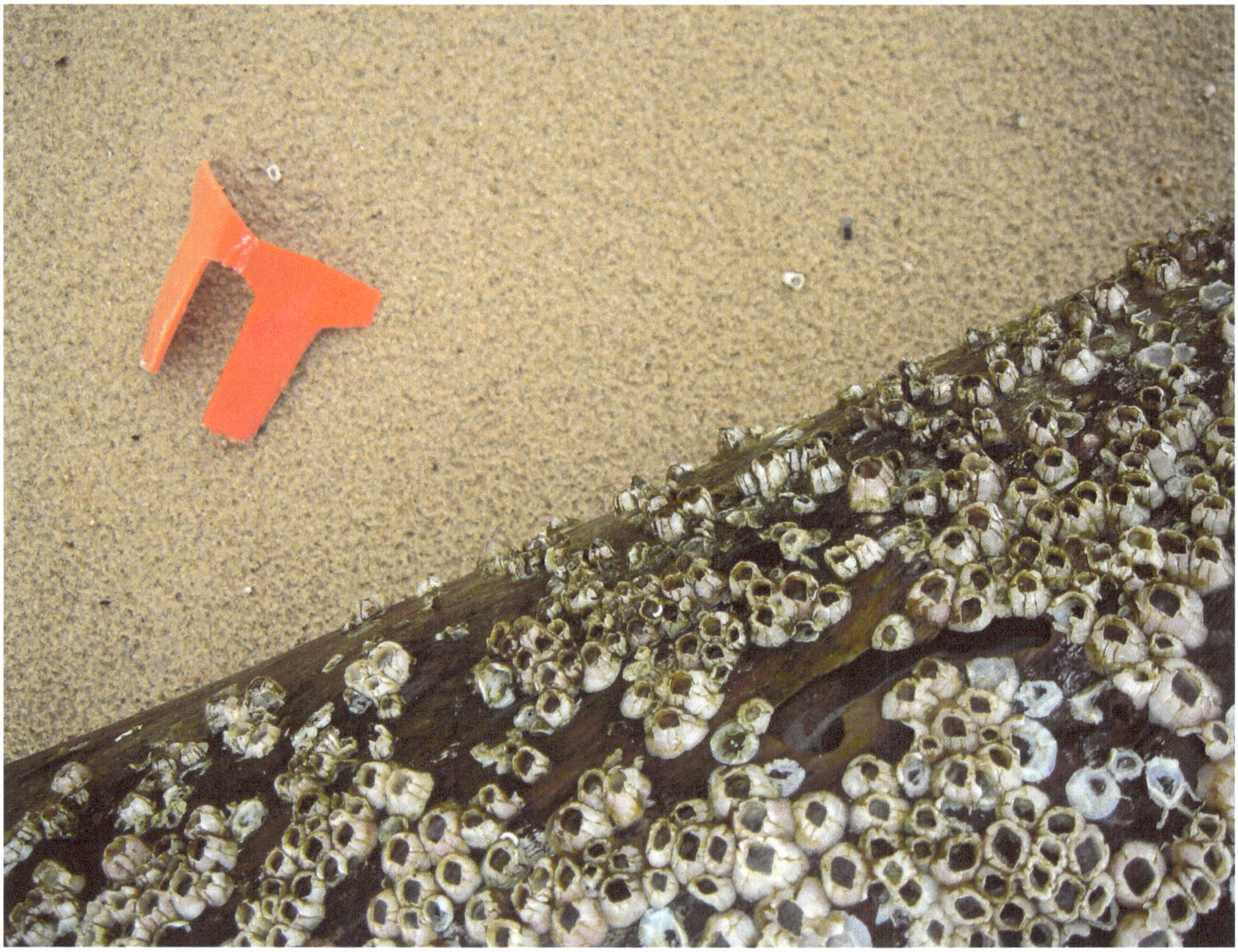

Red plastic and corals

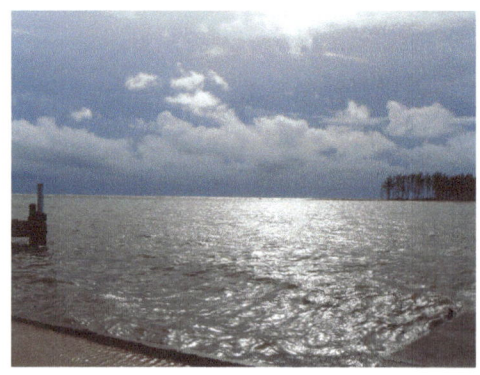

Silent water

you will view the open sea, from where annihilation came. It is here, very close to the Buddha icon's position, where we would have found the heroines of this telling that disastrous early morning.

We had admired the beauty and silence of the sea for quite some time, and had started to walk along what once had been the village main street, when we came across a scene with a house that just about had been cracked by a fishing ship.

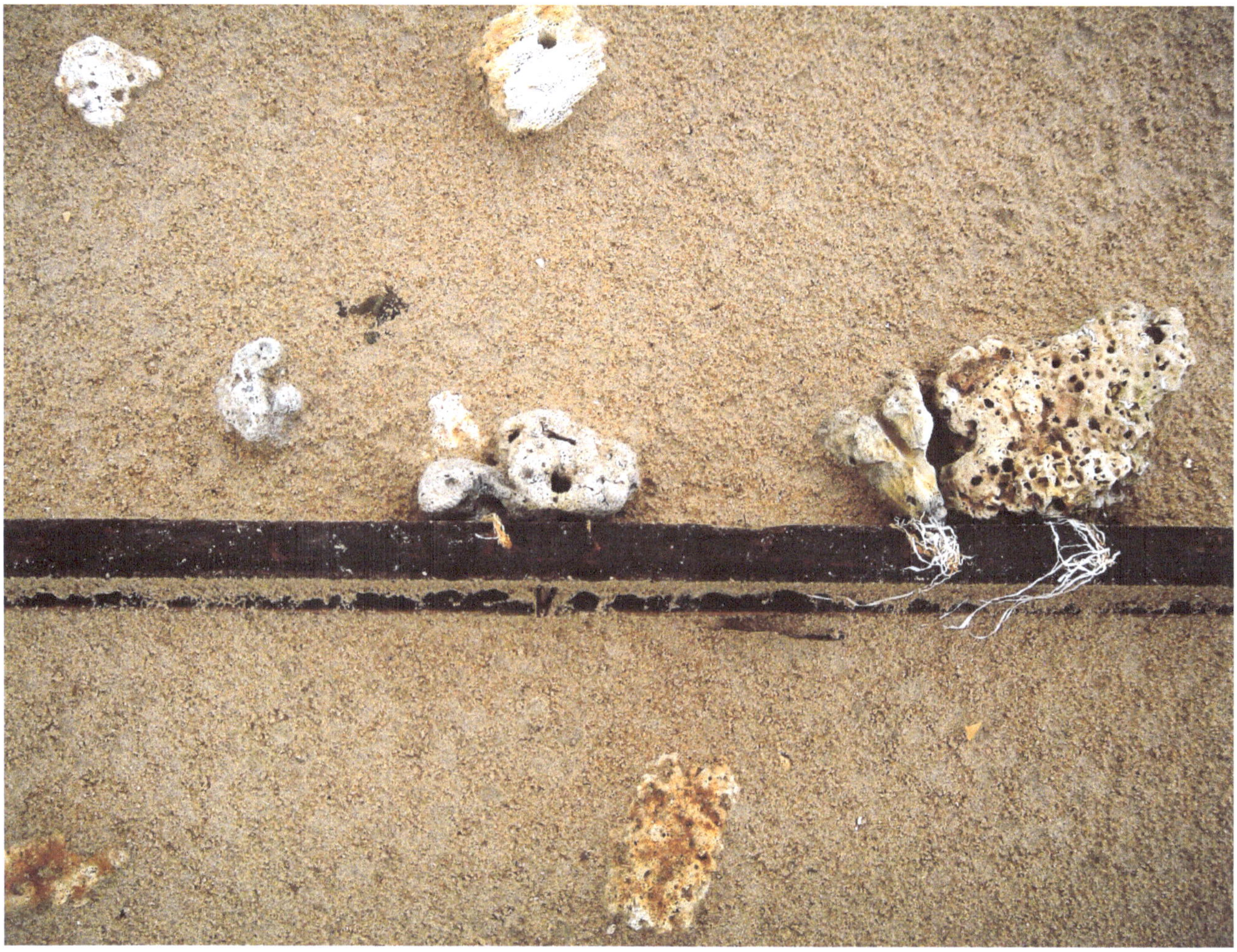

Above and below

I took this picture to record one of the monster's destructive mechanisms. Obviously, the house was there before the tsunami, and must have been thumbs away from breaking into small pieces. I reflected over the ship: would someone make it afloat again?

From the fishing boat, associations went via fish to food. My hunger had become intense. We met a young soldier and

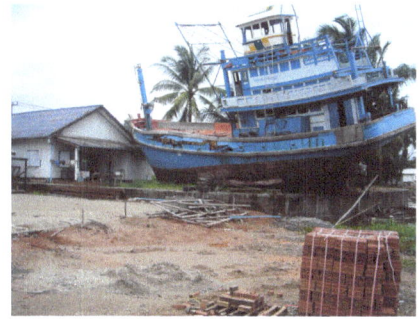

Close to destruction

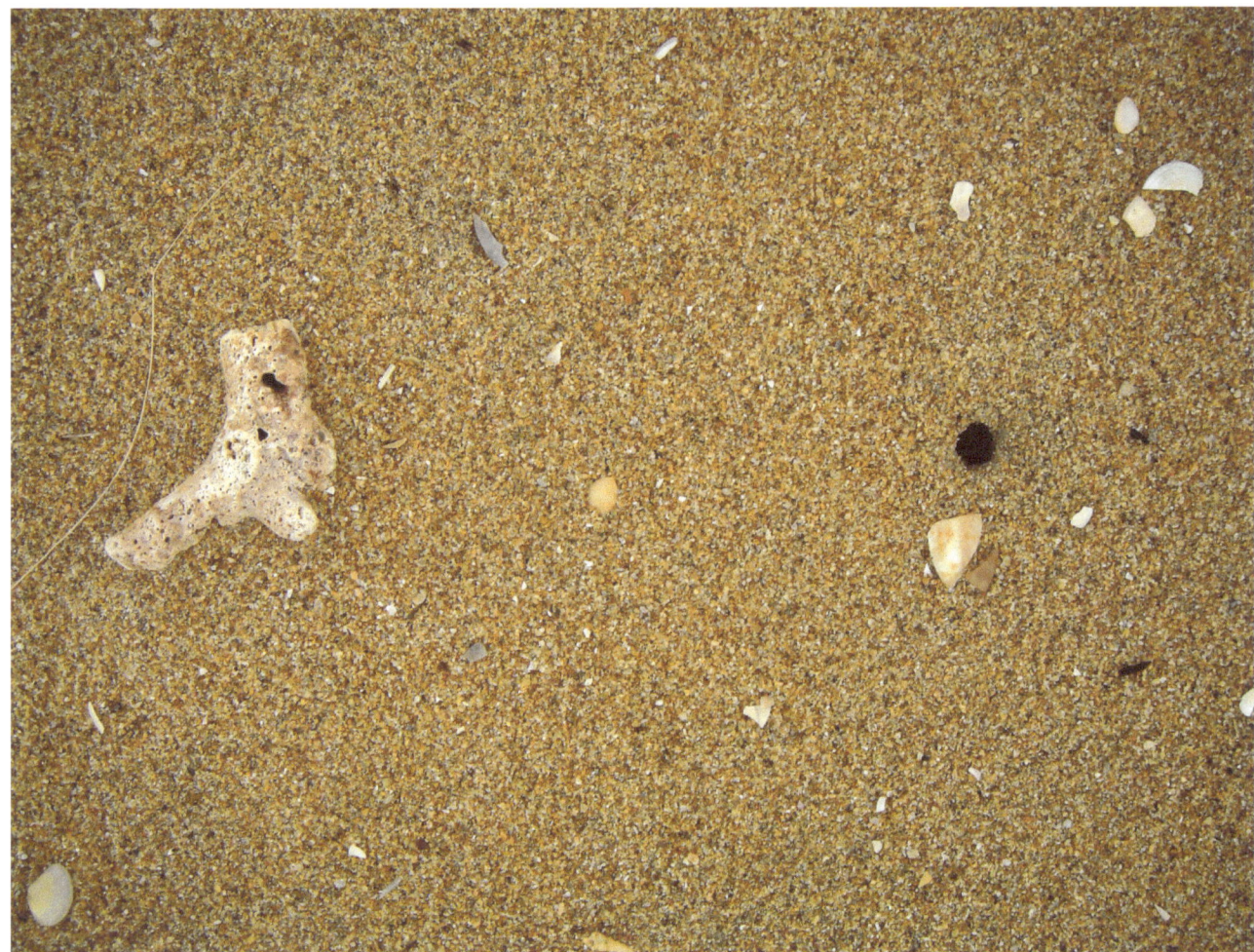

Freedom is to observe and act

asked him about places to eat. Why expect a village with so few people to have a restaurant? We had only met young soldiers this far. A military canteen or something similar would be the best to hope for. We were lucky though. The young man in camouflage uniform told us there was a restaurant up the street. I began to walk very fast.

The restaurant was small and without menu for me to read. As Chiraphorn explained the choices, I noticed there were only ordinary people around, none of those young soldiers. Still, those ordinary people were all men. The men soon began to talk to us, asked who we were.

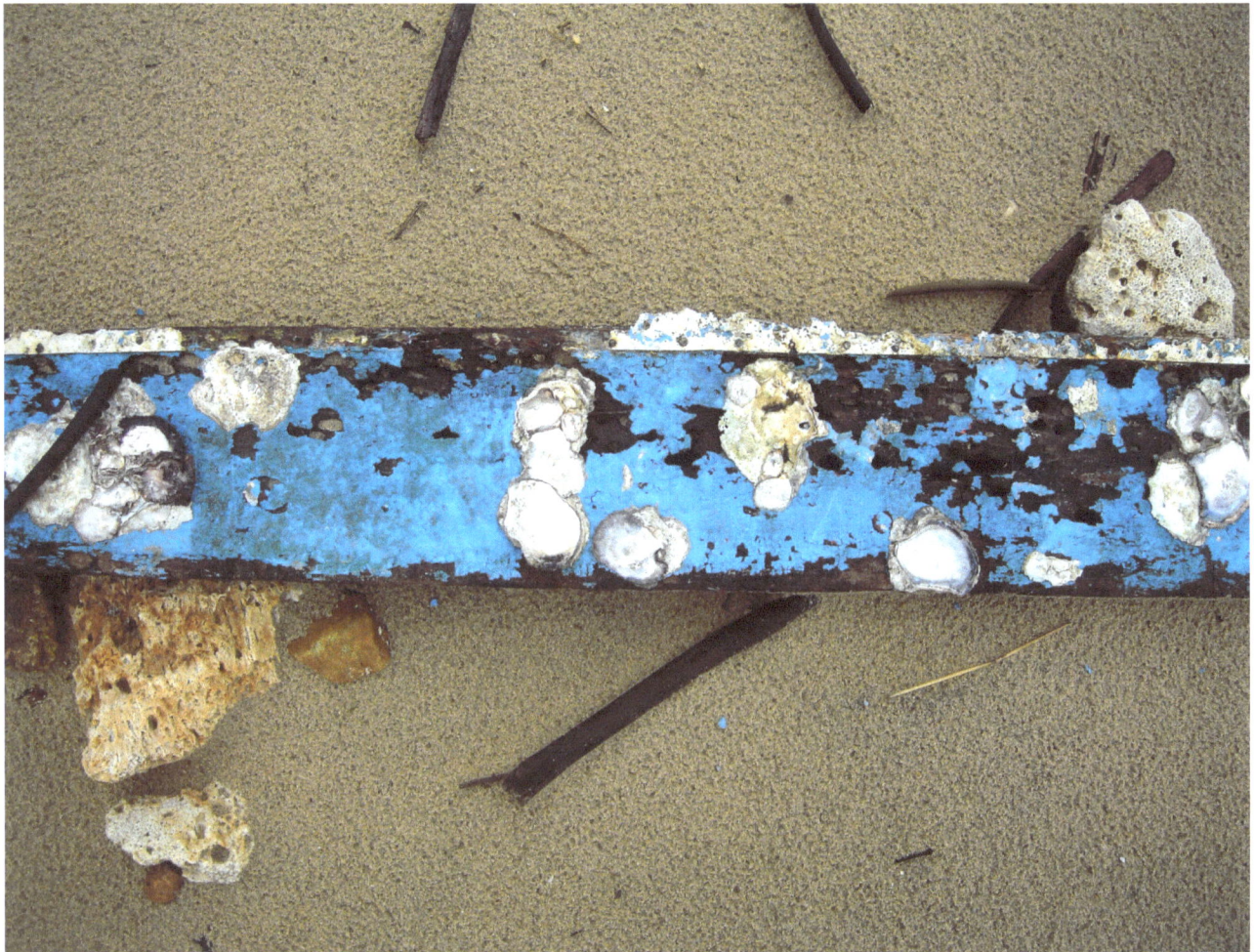

The same blue with more corals

The men began to tell us about how destiny had changed their lives then. They told us every detail of events, as if to cry out, *Here I am big world, please listen to me! I need to tell my story also to you. You too must see and understand what happened to us.* While we listened to these stories I noticed some children. A girl and boy in late childhood helped to clean the tables and serve food. Then there was another little girl, just a toddler. She looked very sweet and smiled to everyone. Obviously, there was something more to her however. The men treated her in a way out of the ordinary, as a treasures gemstone. I thought, *The little girl must be a princess.*

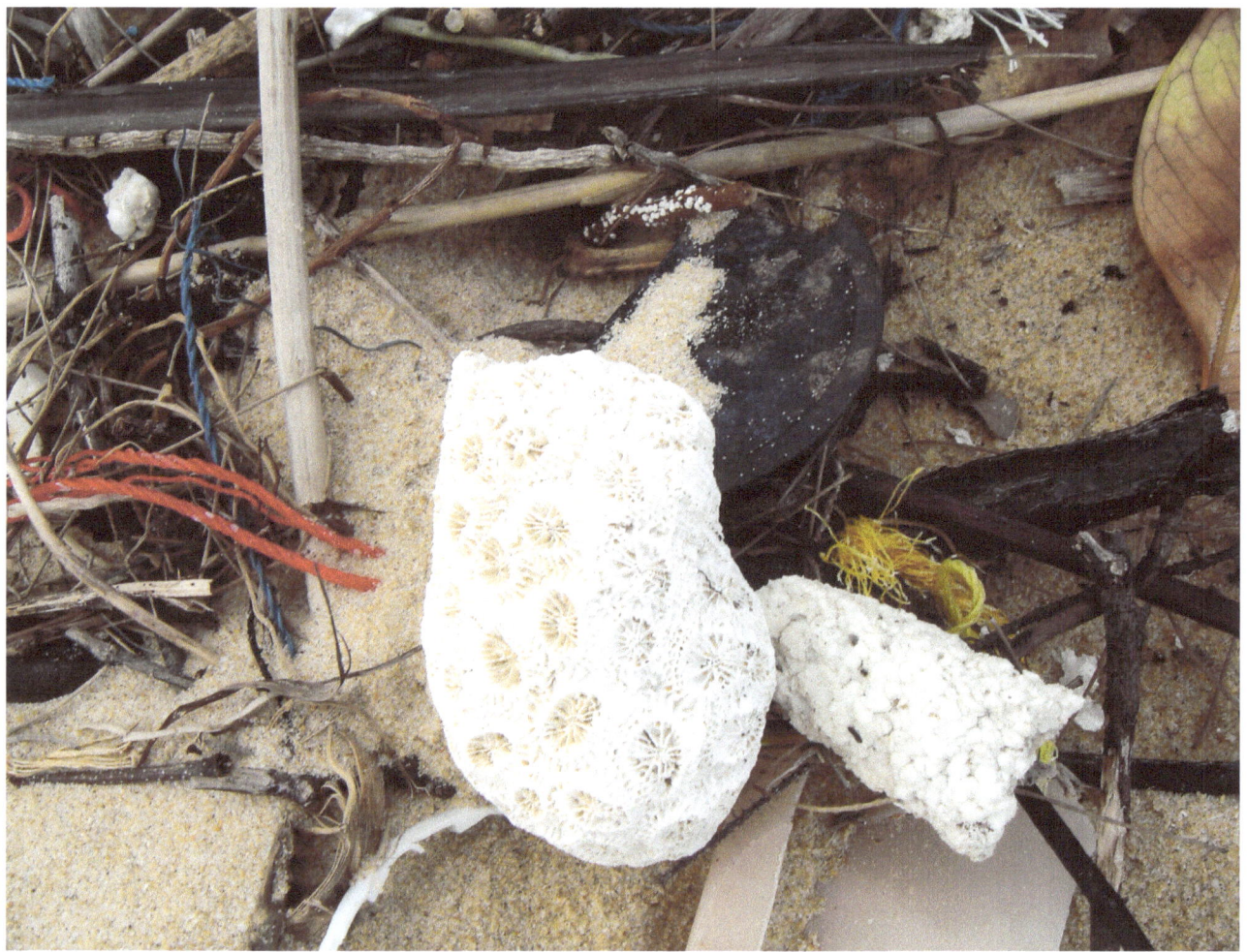

Order gives space

We listened to the men's accounts of the disaster. They were fishermen and had been on their ships far out in the sea when the wave hit their village, killed all what had been of value in life.

We had begun to eat when the restaurant's owners came up to us, the little girl's mother and father. The father told us how they all had been down at the harbour; his wife, children, and he. At the time of the disaster, they had had a little bamboo street-kitchen only, and had just started to make breakfast for the fishermen's return. To explain details clearer, he found a notepad and made a drawing.

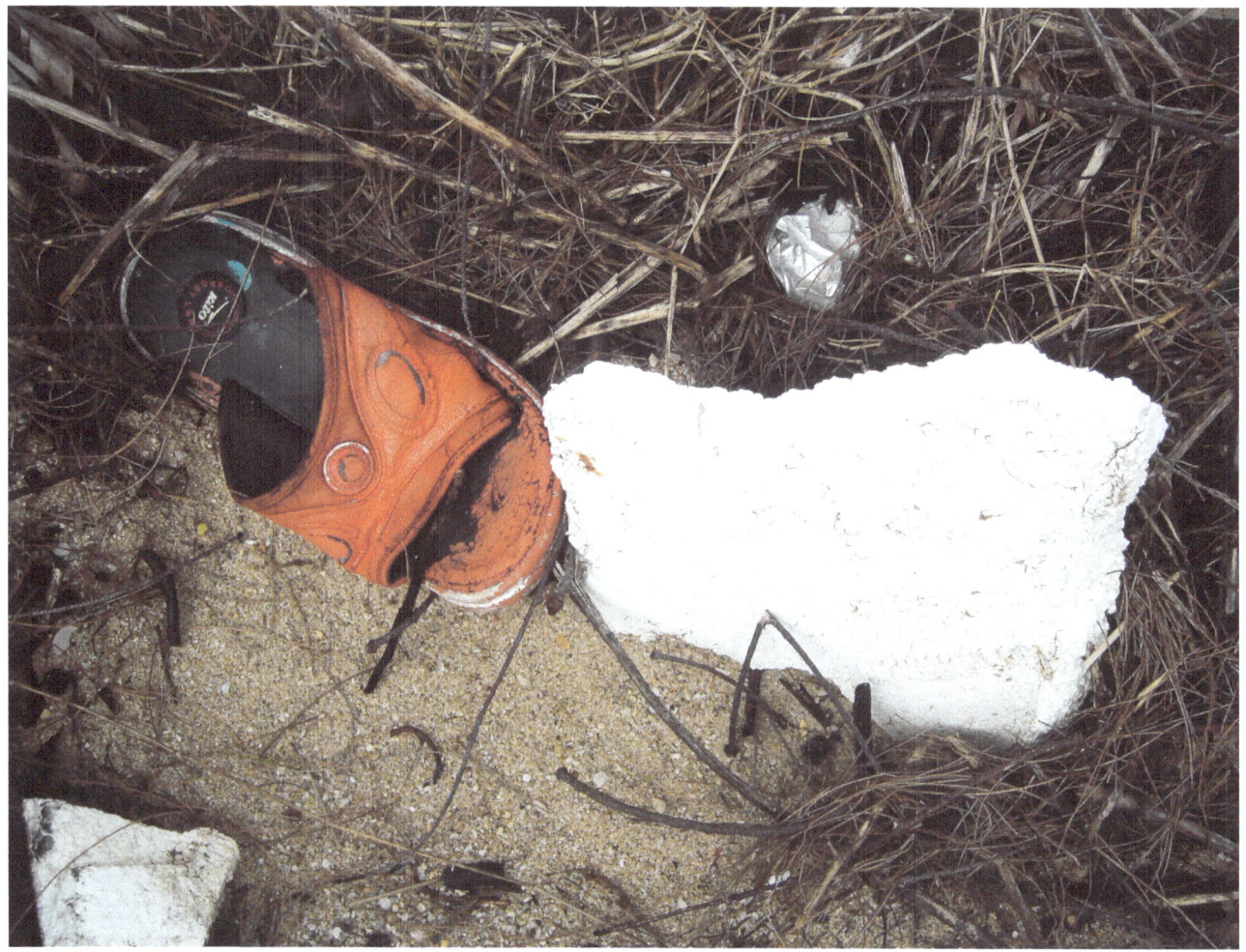

You are all alone in a brutal world

And it was from that drawing I concluded, "It must have been right where the Buddha icon stands now." It was by the Buddha, with that beautiful view of the sea, where the little kitchen, his wife and their little daughter had been that early morning.

The father was some hundreds yards up the road to dispose garbage when the tsunami came. Just by coincidence he had turned towards the sea. He first saw his sister-in-law by the pier and his little son who she was looking after. The little boy innocently toddled a few steps before her. The father smiled happily, before he looked further out over the sea and saw the monster emerge.

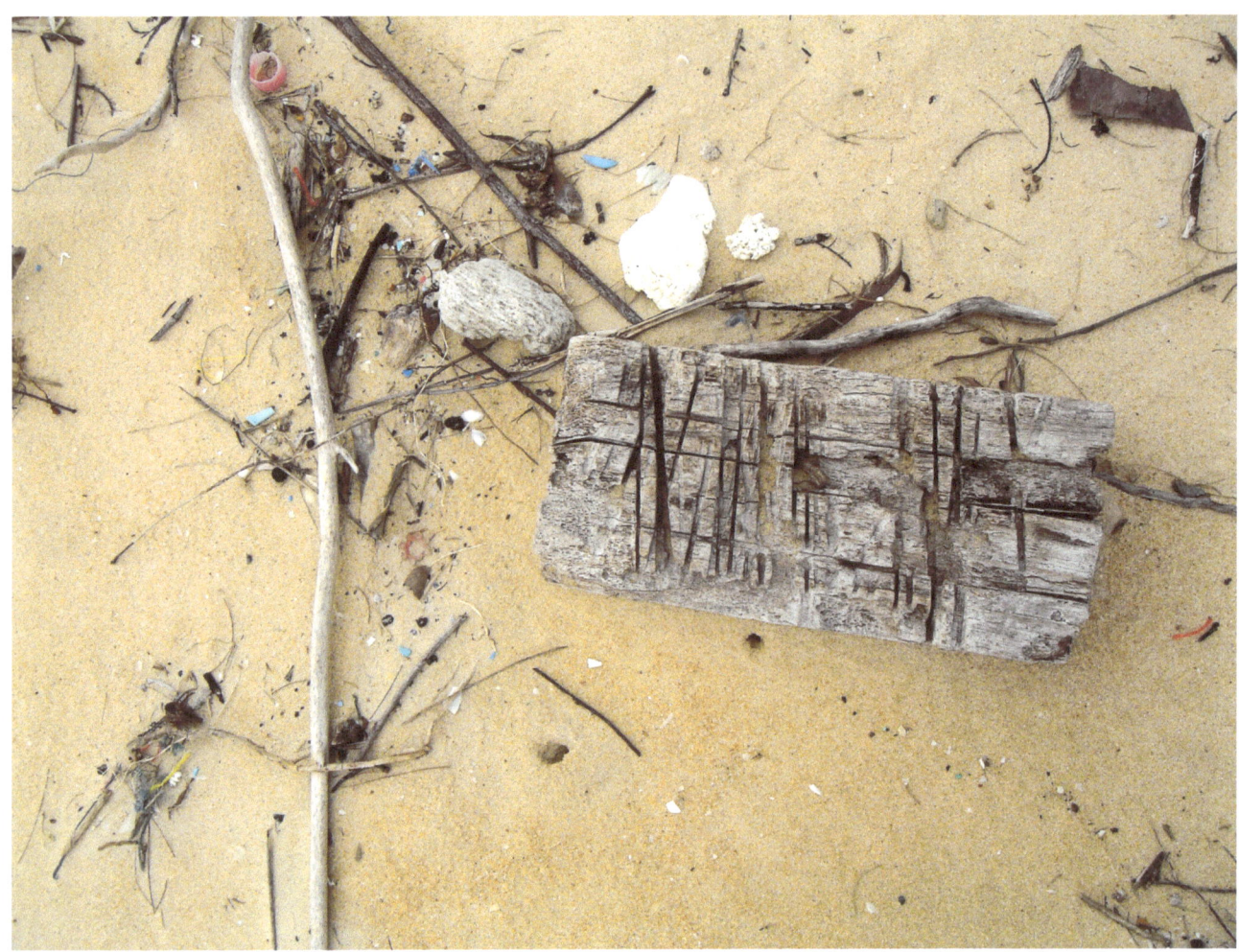

Hardened steel against wood

The growing malevolent monster emerged like a huge wall behind the innocent two down by the pier. The father cried out a warning, and for seconds he actually ran towards them. Then he understood the futility of that running. It was all too late. The monster would only engulf also him. As by instinct he turned back to the garbage heap and grabbed the handle of the empty cooking-oil container he had just thrown there. When he again turned towards the sea, he held it in a tight grip, and took a deep breath. There was nothing to run towards any longer, no sister-in-law and no little boy.

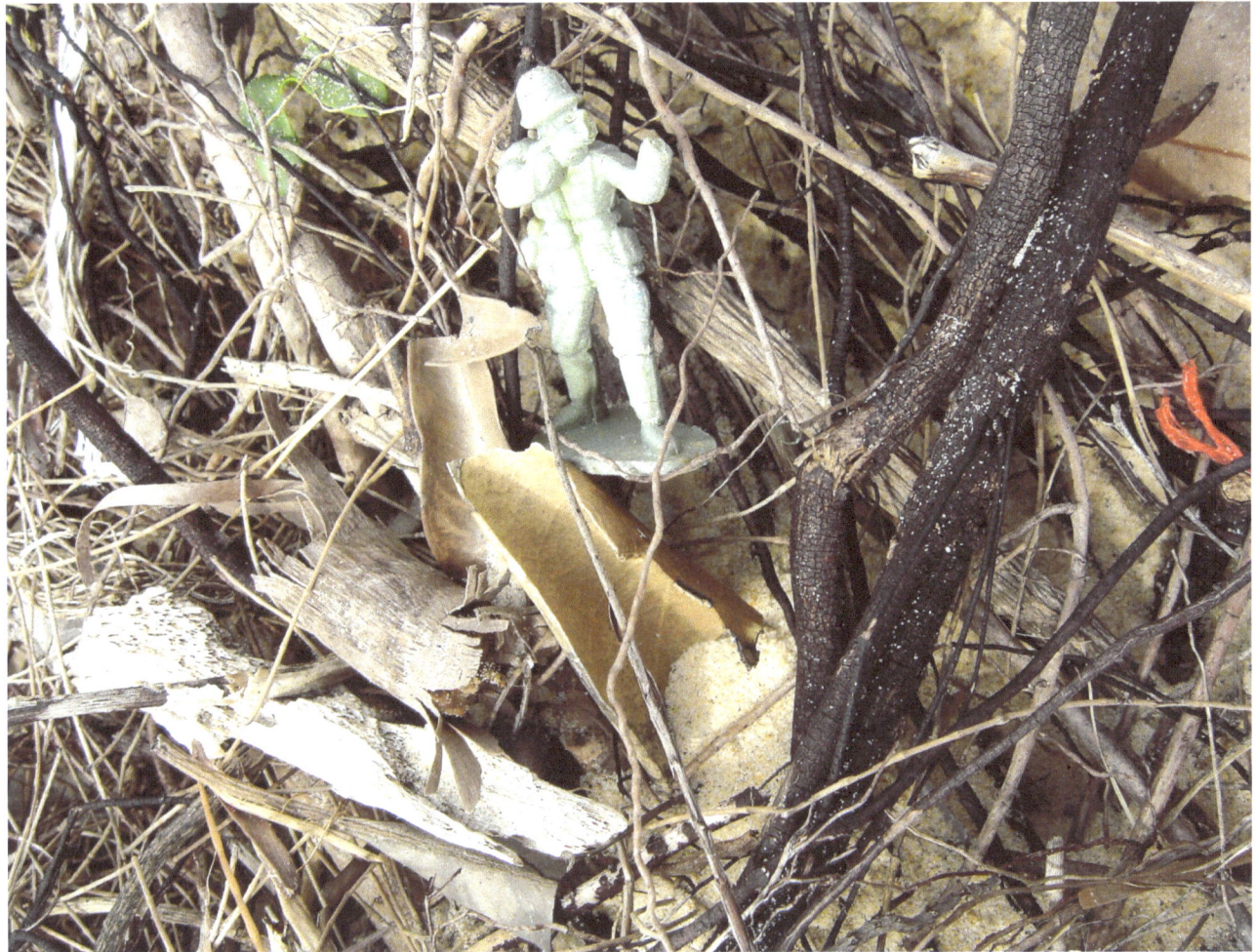

Only a silent mind know what is love

There was nothing more to see any longer. His two children, his wife and sister-in-law were not there. He stood paralysed for a second or so, staring into the huge wall of water. He witnessed the nothing. Water engulfed the whole scene with everything that once had been there.

He did not remember much after that, but only an inner vision of his innocent son toddling on the pier, a surge of dirty water around him, and an immense sorrow. When he woke up a lot later, he found himself stranded on a little island, far away off the coast. The being conscious again and the waiting in uncertainty produced turmoil of agonies, "What had happened to my children,

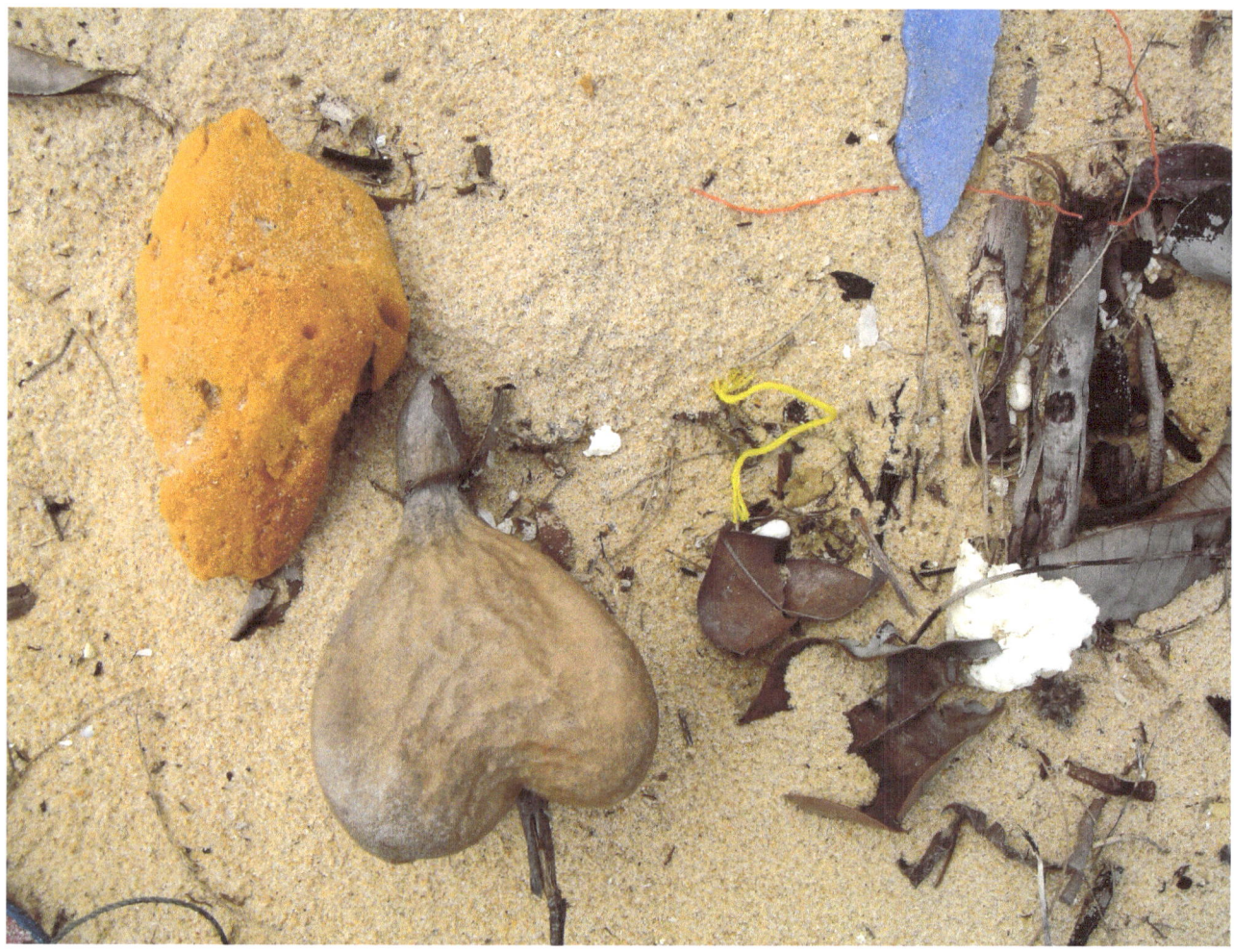

Truth is alive and has no path

wife, and all my others?" There were many such hours of waiting before he finally was rescued, "It started in early morning, but nightfall was already there when our little group of island survivors came to the provisional assembly station."

The mother began her tale now. She made food in her little bamboo-pier-kitchen when an unusual sound came to her mind, "Most likely the surge from water being sucked away from the shore." Also the mother turned around. She did not see her little boy though, but only witnessed the huge giant wave approach.

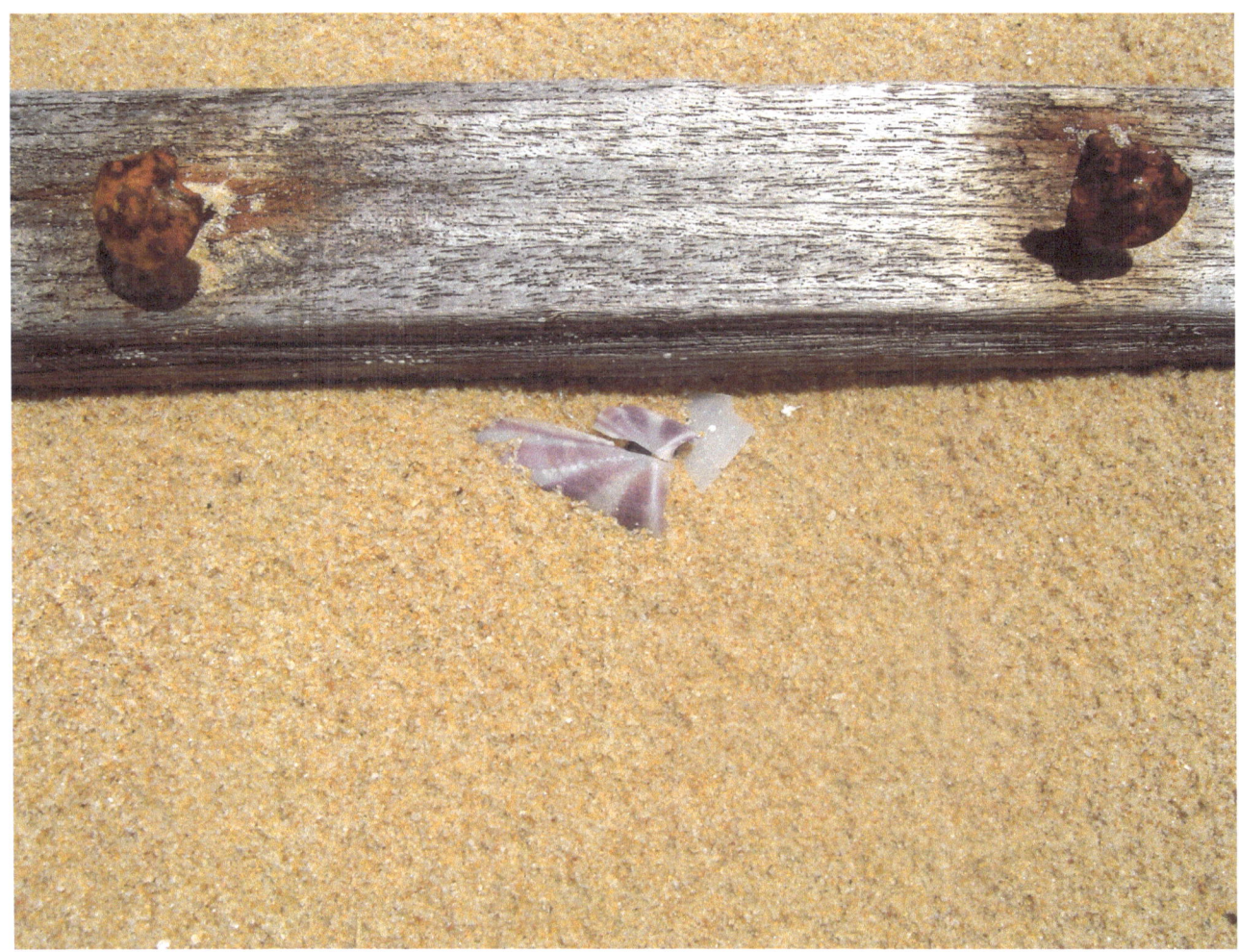

Nails, wood, and a broken shell

By impulse she instantly grabbed her little daughter's arm, "I am so very happy she was that close to me." With her other arm, the mother embraced a bamboo pole from the street-kitchen's construction. Then she neither, just as her husband, remembered anything of what happened for a long time to come. The spell of panic did not leave her for hours. And the first she really recalls is how people forced her to let go the tight grip around her little daughter's body. It was a grip of panic. Too frightened that her little daughter was dead already, the mother had not dared even to glance at her little girl.

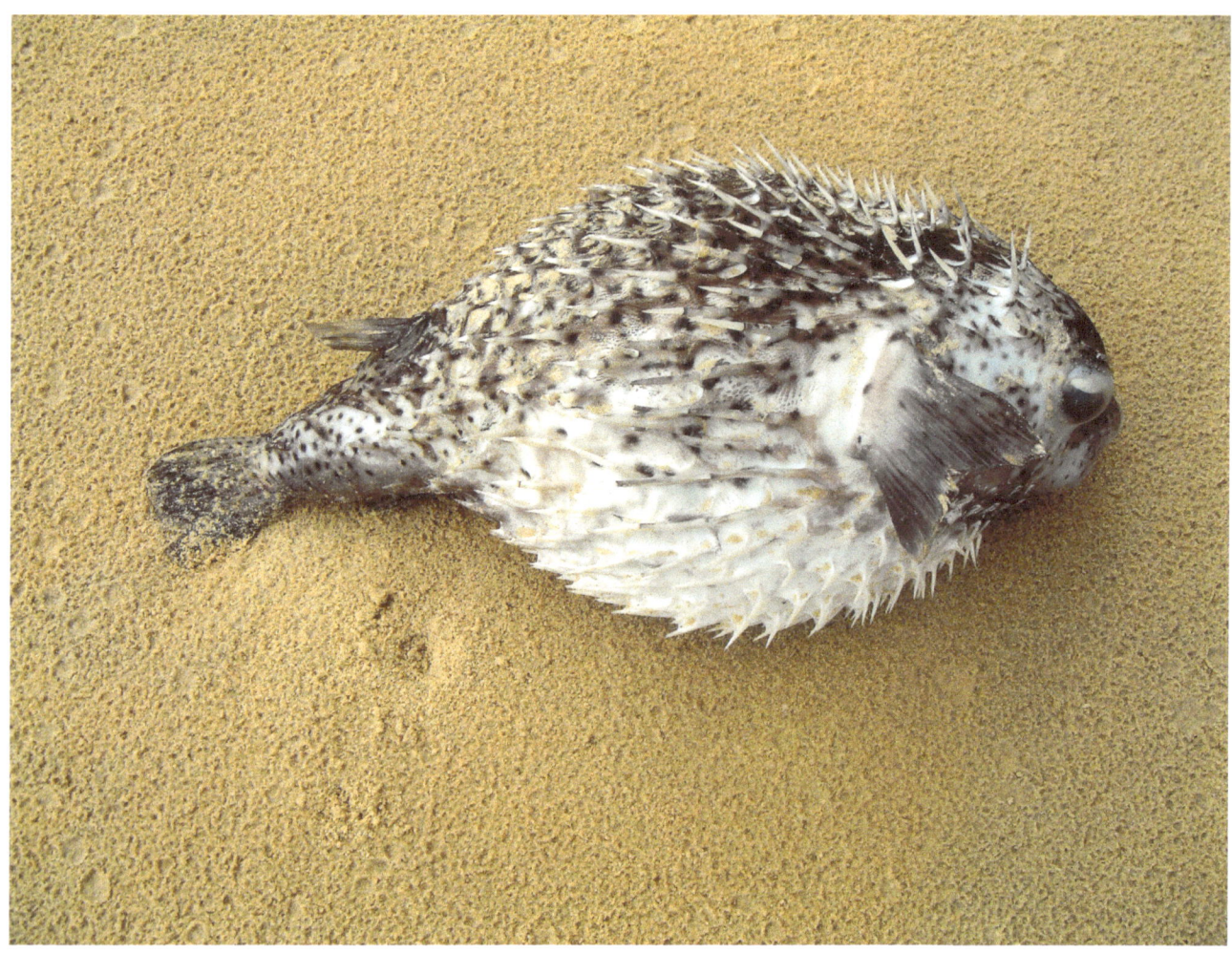

A dead fish

How could it be possible? How could the little girl have survived that agony of water? Yes, the girl was alive! She was not at all hurt. Still, the mother could not take into her the wonder. The little boy was not there, neither the father, nor all those others belonging to her kin.

There were no scenes of joy when the father turned up later that very same evening. The father's eyes expressed only sorrow, "I knew we had to go searching for the boy."

The people around forced the mother to remain in the camp that night, but the father refused to listen. Nobody could stop him and he began his search for the boy long before sunrise.

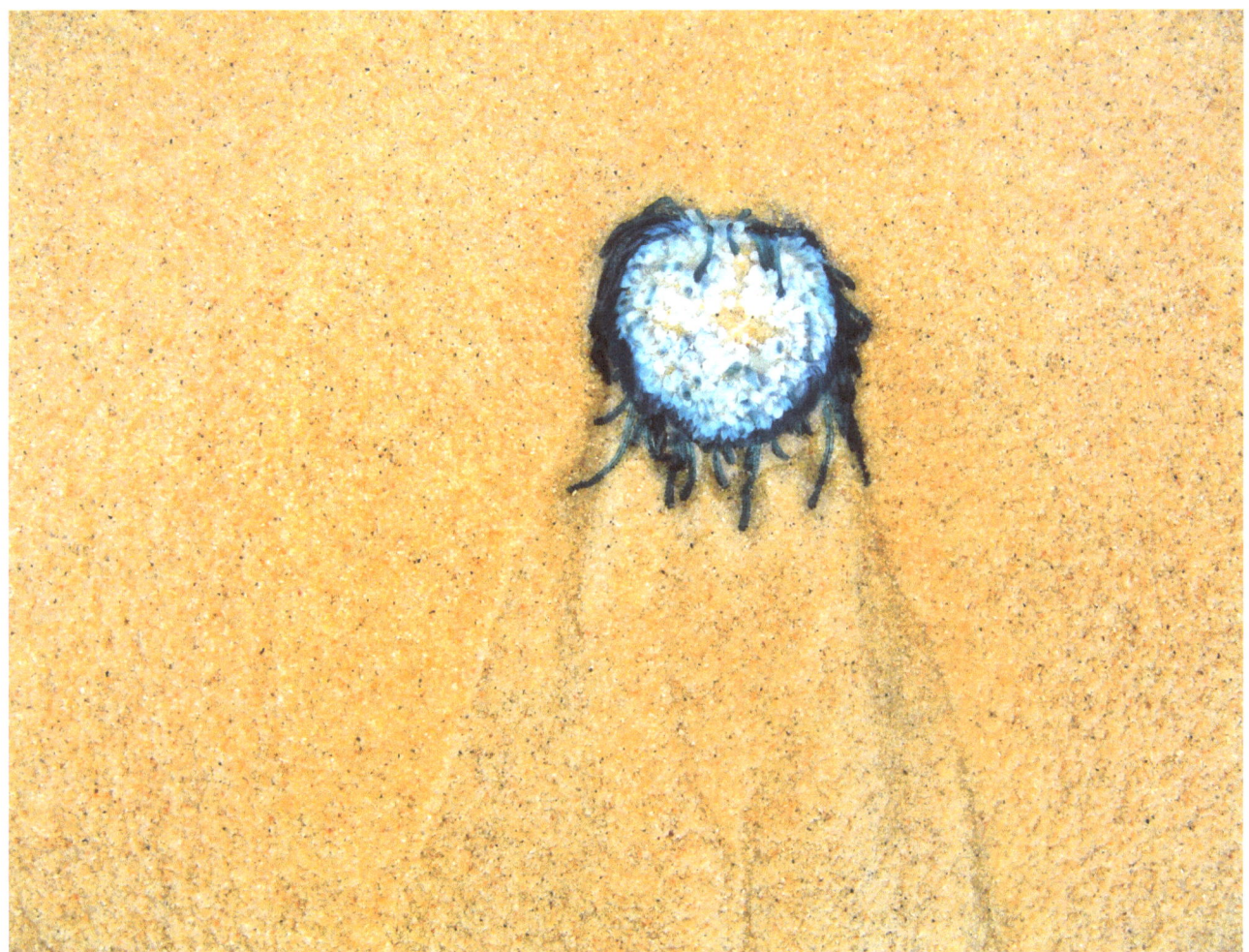

The sand is still wet

The father cannot recall when he slept or when he was awake, if he at all was sleeping. His search continued day and night until, many days later, he fell down by pure exhaustion. At that time however, the military had taken control and simply ordered him to stop his search. He had no alternative but to obey. He then rested one day before he joined the organized rescue-scheme.

Tears came from the father's eyes when he told about what his rescue-team came across in their search for survivors, from scenes of prior human agonies and from people's meetings with their already dead family or friends.

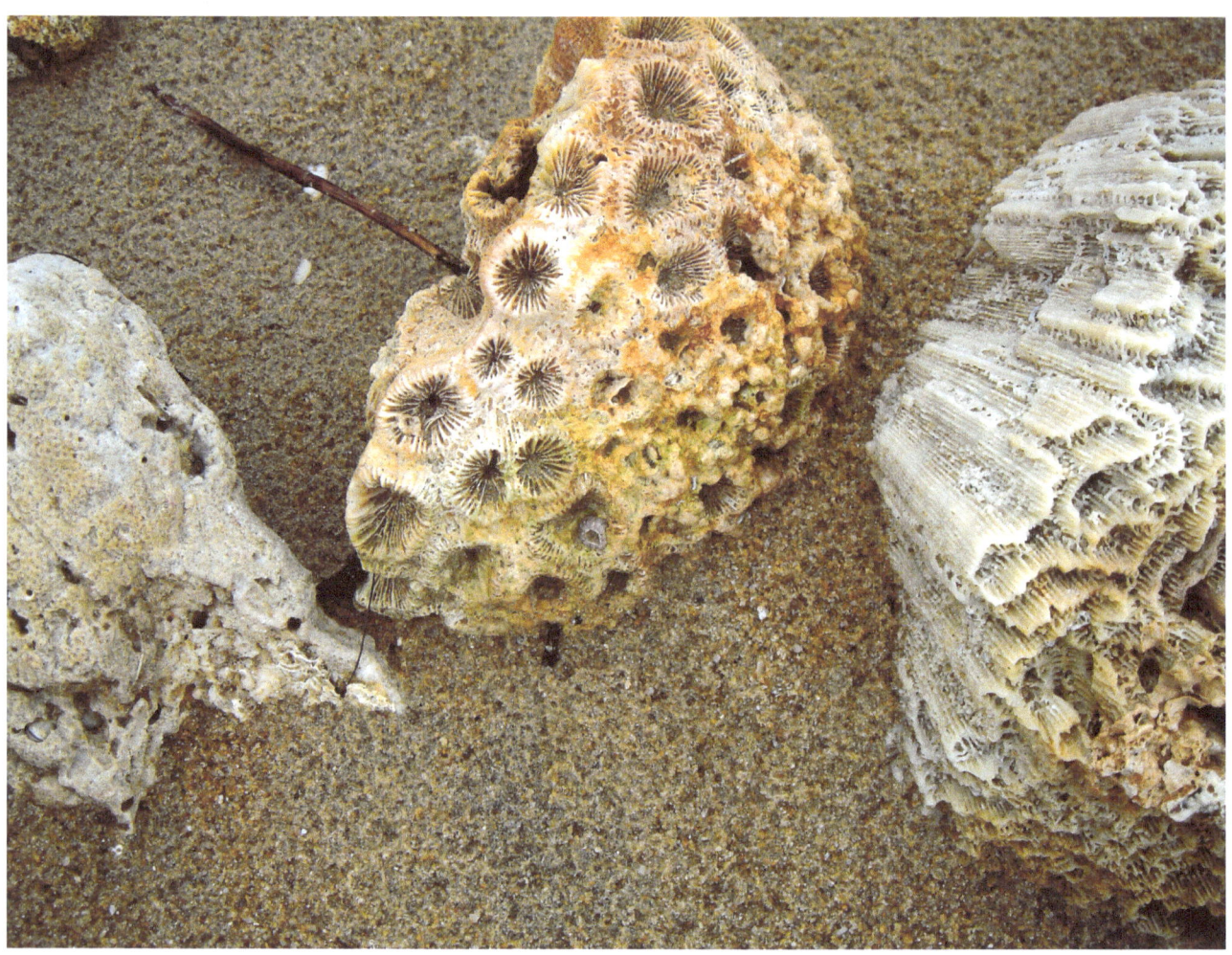

Dead corals and a tiny stick

The father's team never came across his own son or his sister-in-law. Some of his relatives where found and buried. Most of them had just disappeared though, without anyone ever coming to see them. At that time and still in present days, there were rumors around. Hearsays about somebody, somewhere, that had seen someone who looked just like the little boy. The father still thoroughly investigates all such possible suggestions, maybe a little less now than before. Until this day, all those rumours had all been without substance.

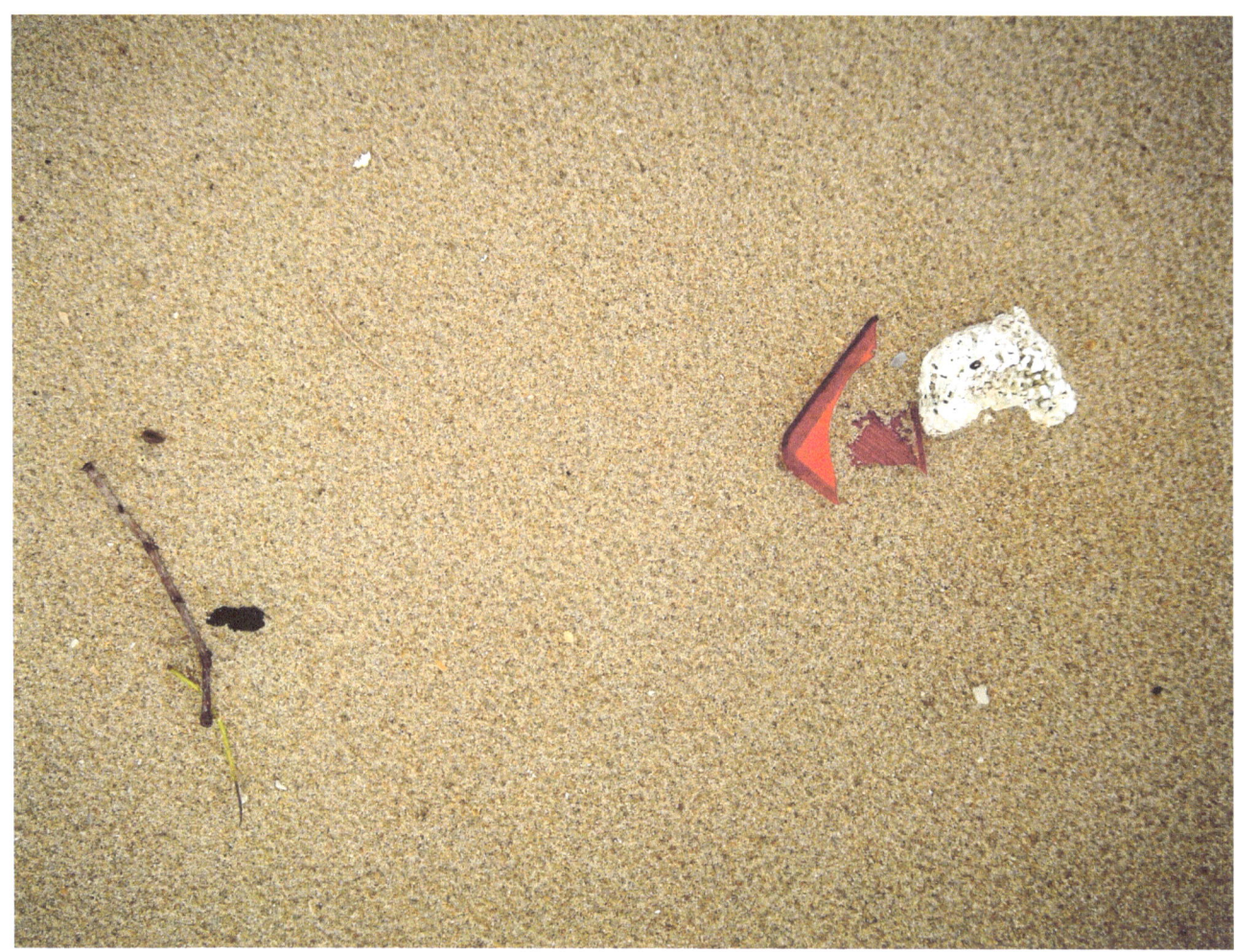

Love is a state of being

At those early days after the tsunami every day became like a rollercoaster of dread and hope. It was about other's grief and losses, or about other's paranoid, at least half-delusive states created by rumours: suggesting for example that someone already had kidnapped the little boy, anyone who had lost his own son. "Why could that not happen, when the wish to get back the ones lost was so strong? Why couldn't anyone's son do?" How many times had the mother and father not wished that the next living child, when seen from some distance, should turn out to be their own little son?

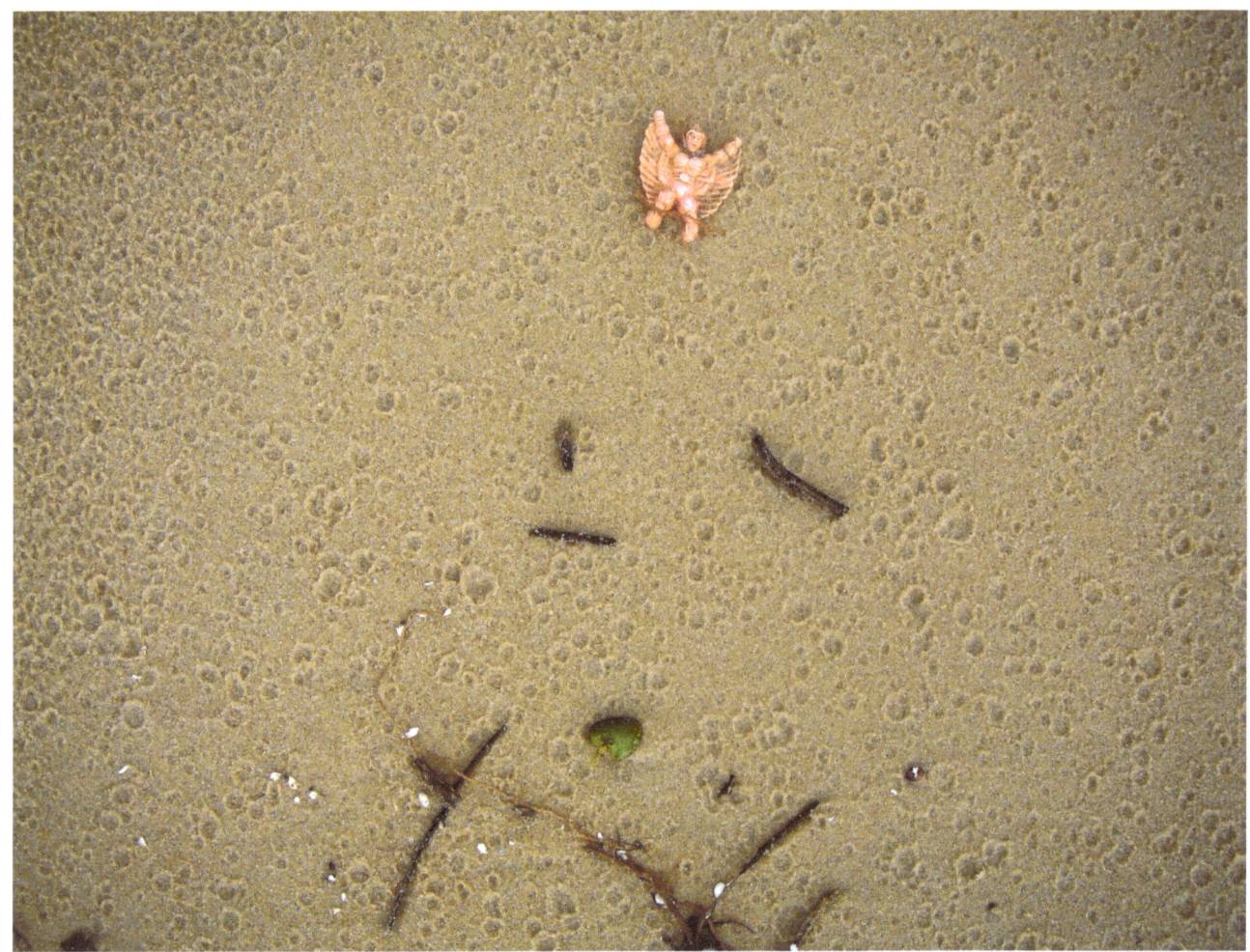

It recently rained

The agony of loss continued for both the mother and father. The loss of the little boy might have helped them in one way though. They tried so hard to find their little son that they did not take into themselves the unlimited tragedy they were part of, the entire misery around them. The little son occupied their entire mind and thus became as a symbol for those others who were gone: the sister-in-law, parents, grandparents, neighbours, friends, regular customers by the bamboo-harbour-stand, and the beggar who used to turn up at closing time for some left-over. Those folks were all gone now.

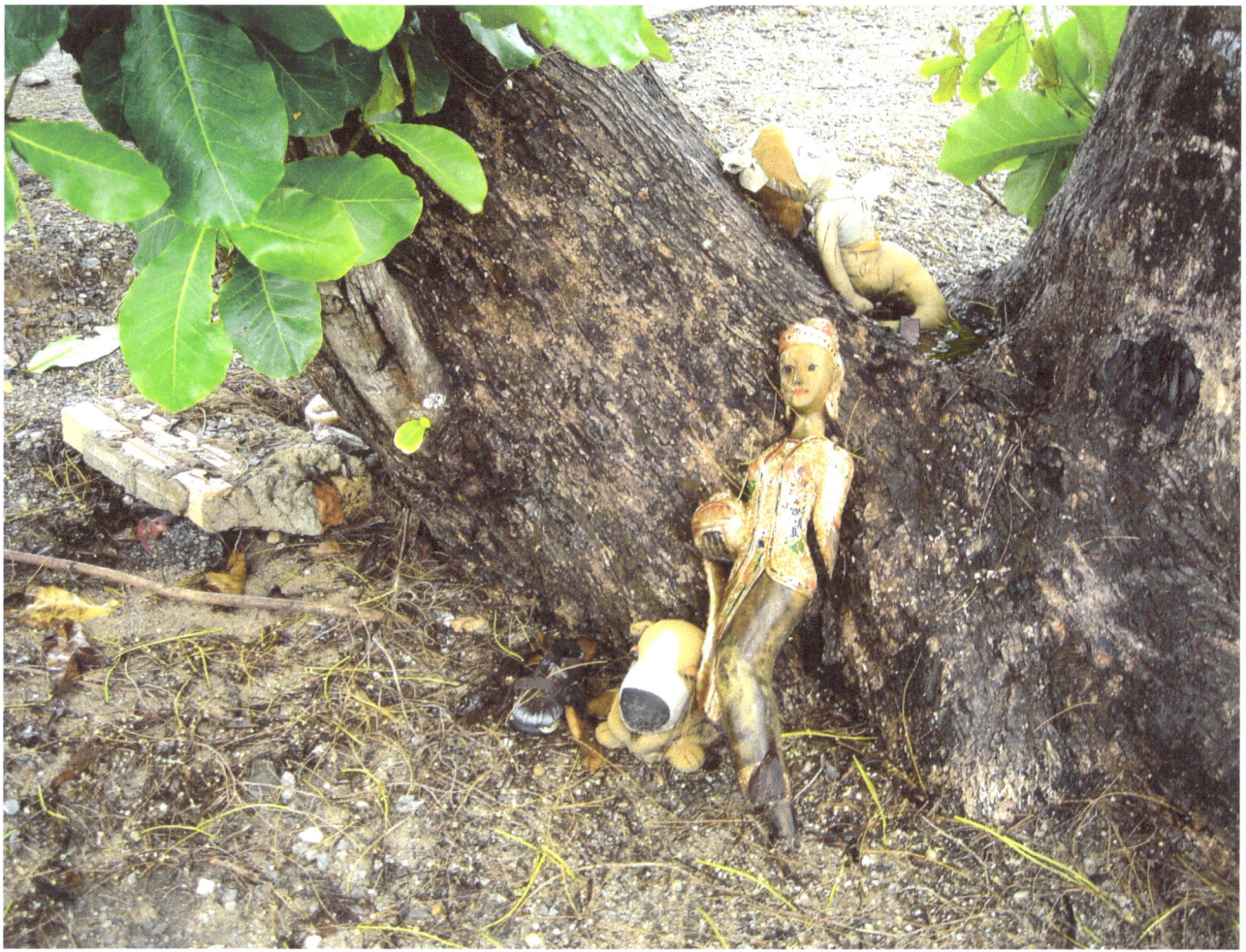

A provisional gravestone

If a monstrous wave destroys an entire village, like this little fishing port, very few of those who were there at the time could have a chance to survive. How is it possible to survive? We already know about the fishermen who almost did not notice the tsunami. This little family was a miracle! Were they just lucky or was it about an inner meaning? When the tsunami came it must have been the very fewest who could hold on to something floating, like an empty plastic cooking-oil container or a large enough piece of bamboo. There were only men at the restaurant because this act of God had taken away all that was dear to them. The monster had left everyone of them alone, and their hearts empty.

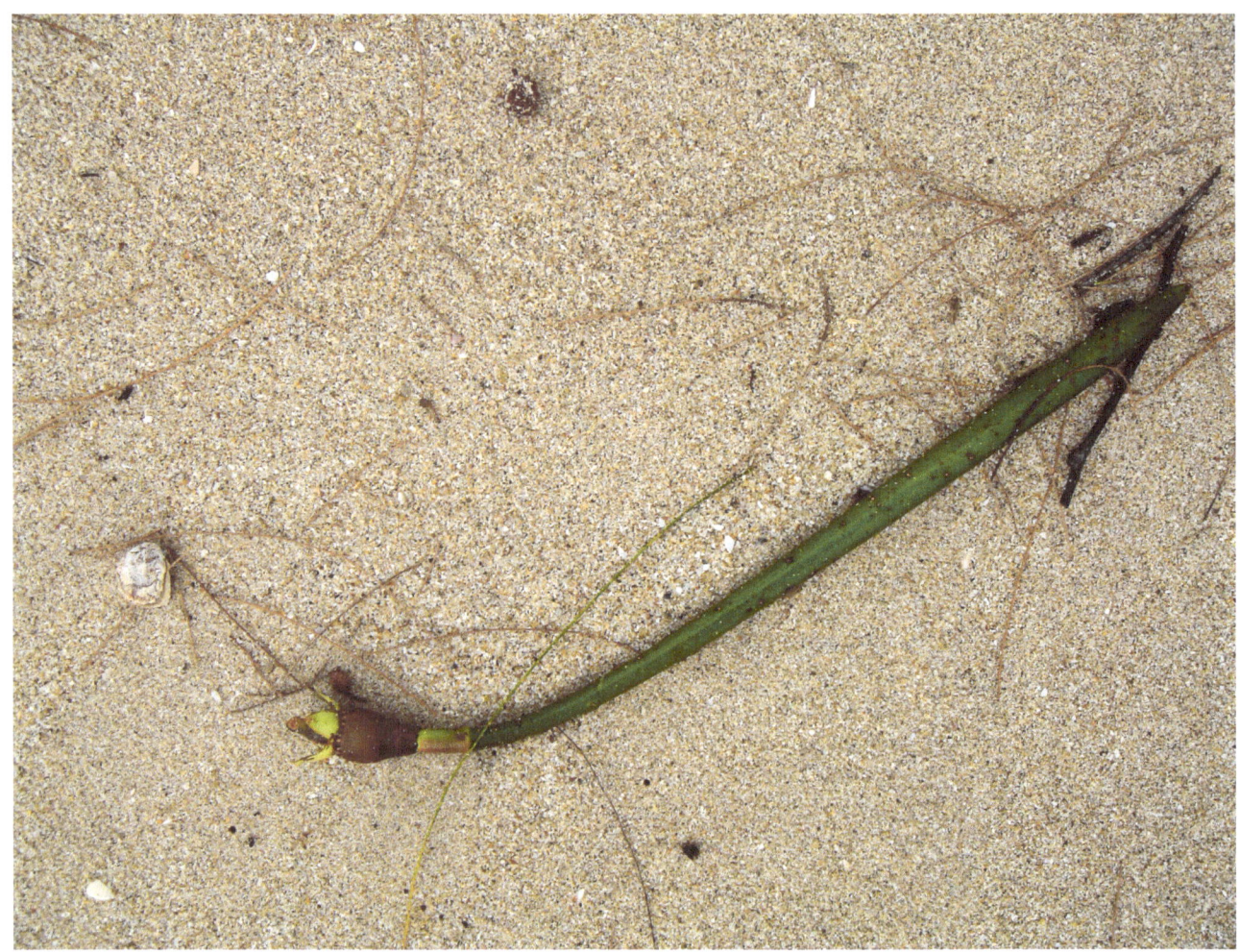

Observe anger without opposing it

These men had been away working on the higher lands or fishing at the see that early morning of the disaster, while the women attended the houses, took care of their children and old folks. Yes, those who had been in the houses were all gone, so were the children and old folks, and almost all the others who had been there in this village, situated far too close to the sea.

I saw the sorrow behind this father's story of survival. I saw sadness even in the drawing he had made. Sadness lured behind facts and figures explaining his survival. An aura of sadness surrounded him. I looked around then and read this same sadness in every other man's face.

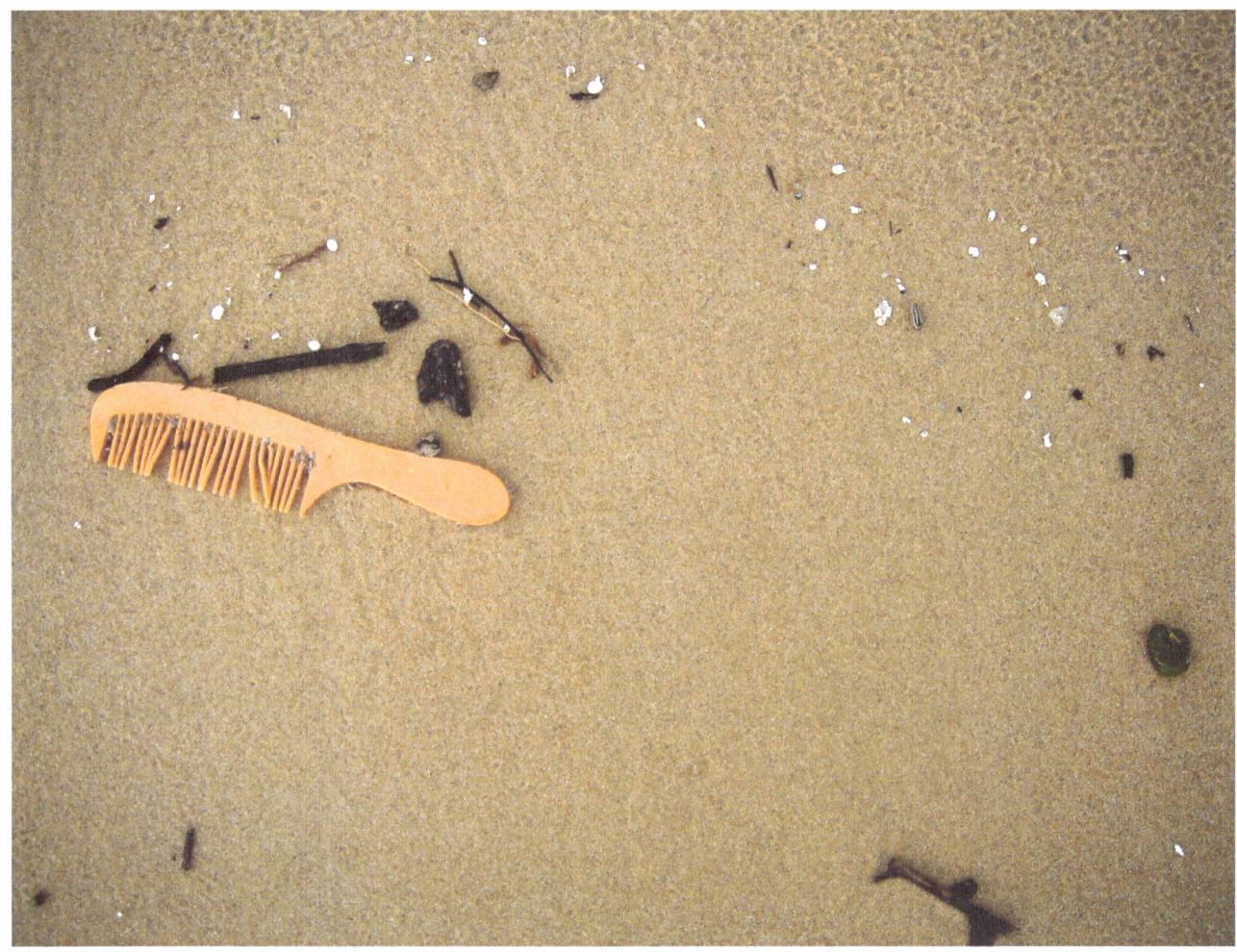

I like the satisfaction of helping others

I had not grasped it before, but suddenly it got into me with all its strength: these folks were without kin. They had lost all what was dear to them.

We got close to the question of what life was meant for. Why do we live? For what is this life good? Such questions are natural when all is lost. What can you do when all is lost? Some ran away, had left the village to find happiness far away from the place they were born and grown up. Other survivors suddenly died or just disappeared. Did they kill themselves or leave in silence? They were no longer here but were now refugees of inner pain, had left despair behind.

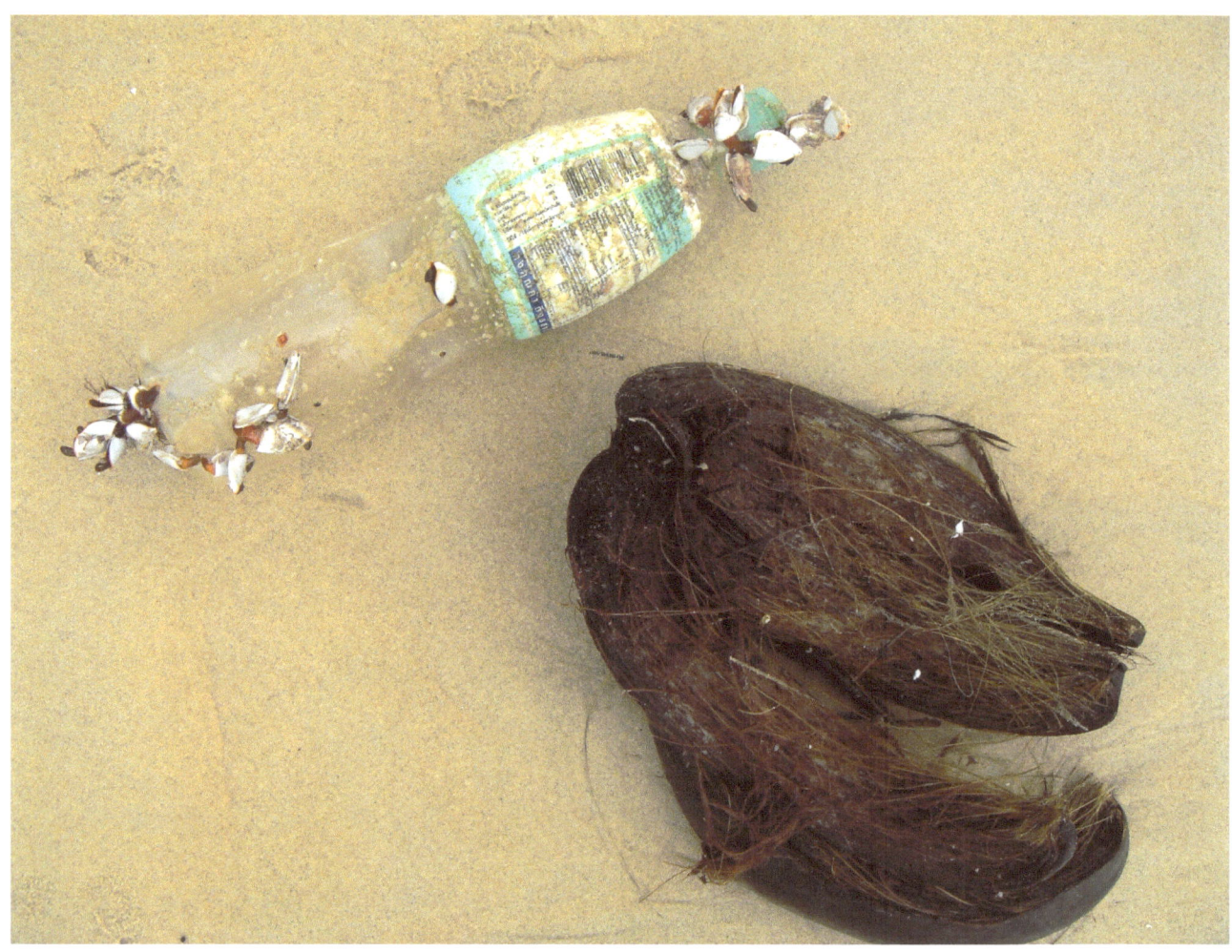

A plastic substrate for life

You must be very brave to face your own total loss of kin and friends along those Indian Ocean beaches, where family connections means everything. In this part of the world, the fewest would understand new concepts of material and social security. You just do not live alone here.

What content does your life have when you must live alone? How can one even talk about such complete despair? Why doubt that so many of those tsunami survivors had died from pure mental exhaustion, from the grief and despair of losing their kin.

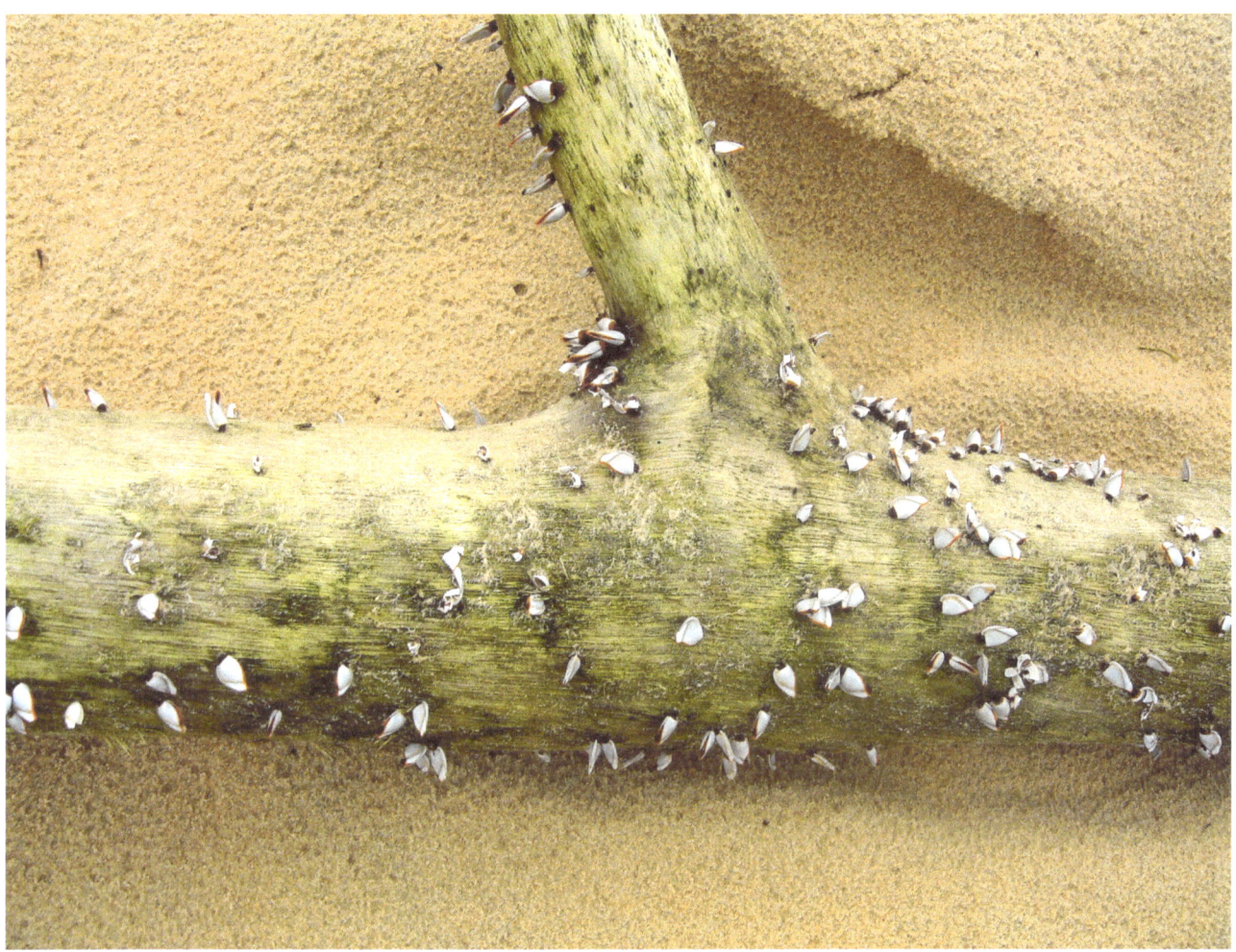

A wooden substrate for life

One of the men said, "How much did I not want to die? I would have been dead if it wasn't for attachment to family and homestead, which now are gone." I asked myself about arguments for being, or not being, in a life devoid of relations. Was there another argument for being alive?

As if reading my wonderings, the same man helped me with the answer, "The only real reason to live is the mere fact that we are alive. We were born into this world and then life itself obliges us to continue to live?" If you die to what you know, as this man had done, then that dying brings innocence and passion. Only dying can show you that life itself is the true reason to be alive.

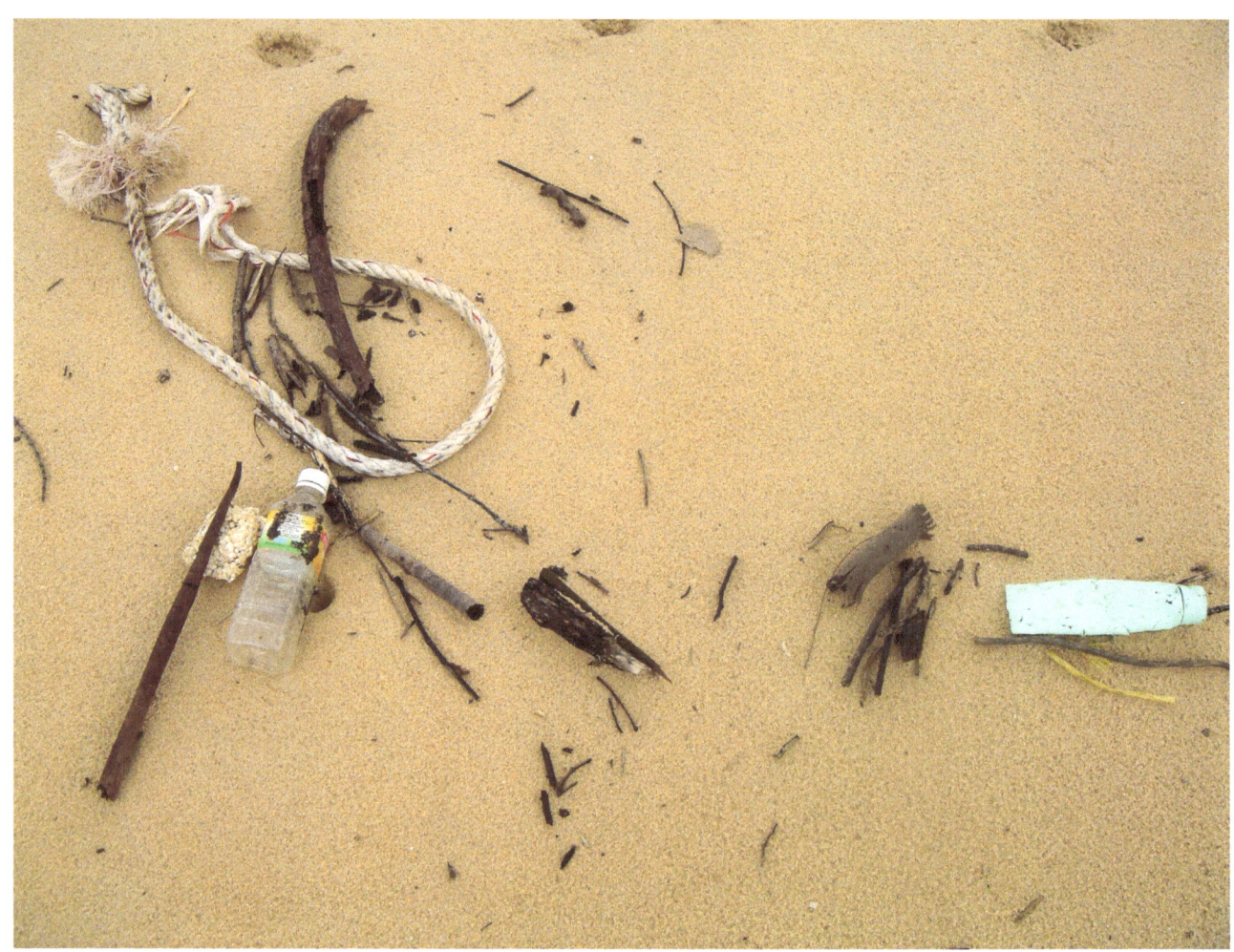

You should not be thought's slave

The girl who had served us the food asked me if I also wanted a beer. I nodded a "yes" and while she went to the refrigerator to get me a Singha beer, the father explained she was a Cambodian orphan. A few months ago, she came here with her brother to stay with them, "The two orphans bring new hope to all of us, as they were our own children, belonged to the village. Although somewhat older, the boy could have been our own vanished little son." I looked at the boy. He had no time to look back though, proudly looking after his little princess sister.

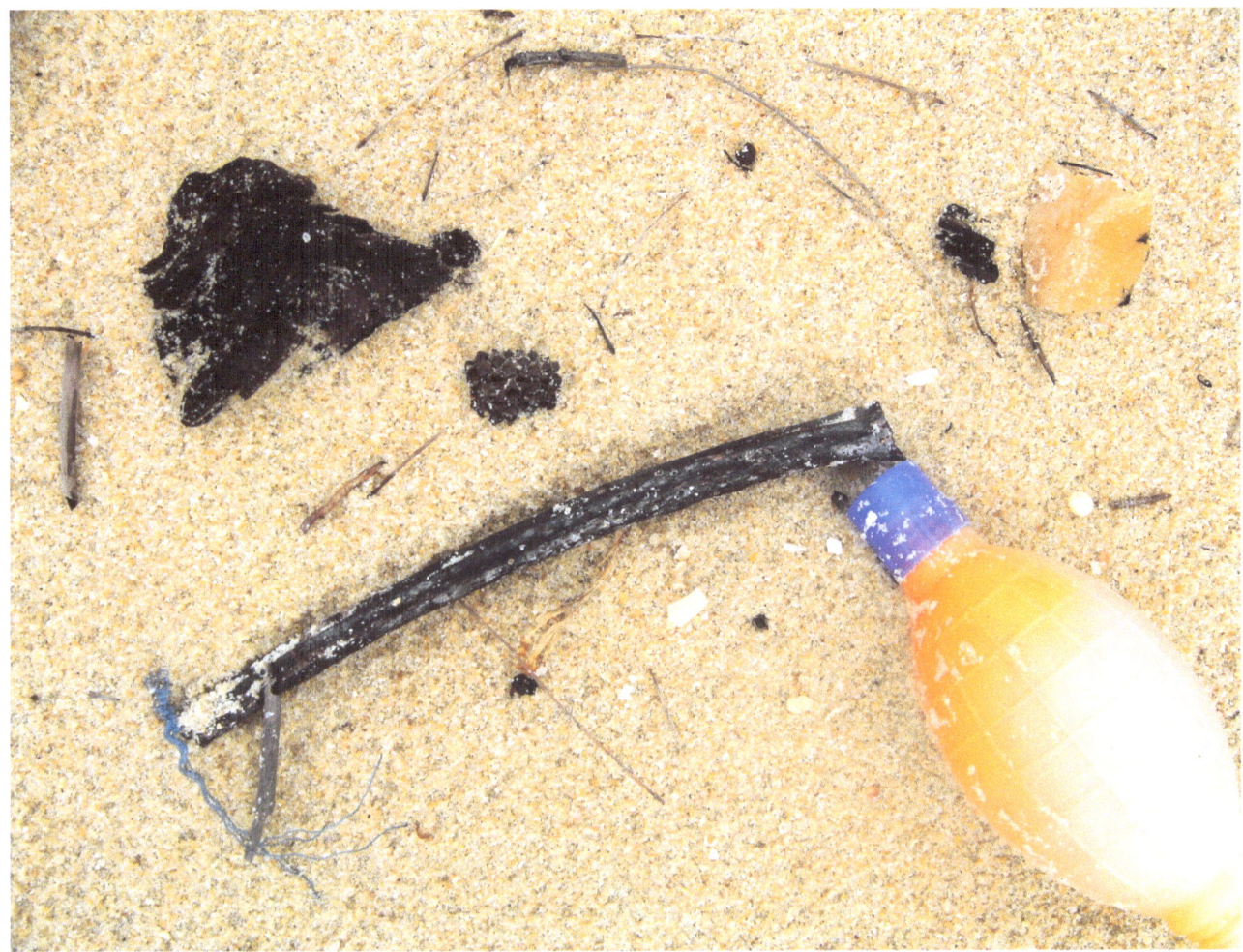

Humility begins where conceit ends

The food was delicious. I love seafood even more than **Filet Mignon steak** or Swedish Meatballs. And all we got to eat was perfectly fresh, including a crispy lettuce. The taste was like in any luxury restaurant. What they served us here had never been close to deep-frozen. Something more added to the flavour however. When we sat there eating, we became part of that same extended family of survivors. I became part of the whole scene around. In that state of closeness, the little girl appeared in a new light. By being there in the centre of everyone's attention, she became a flower of beauty and loveliness. She was everyone's little girl, and everyone's hope.

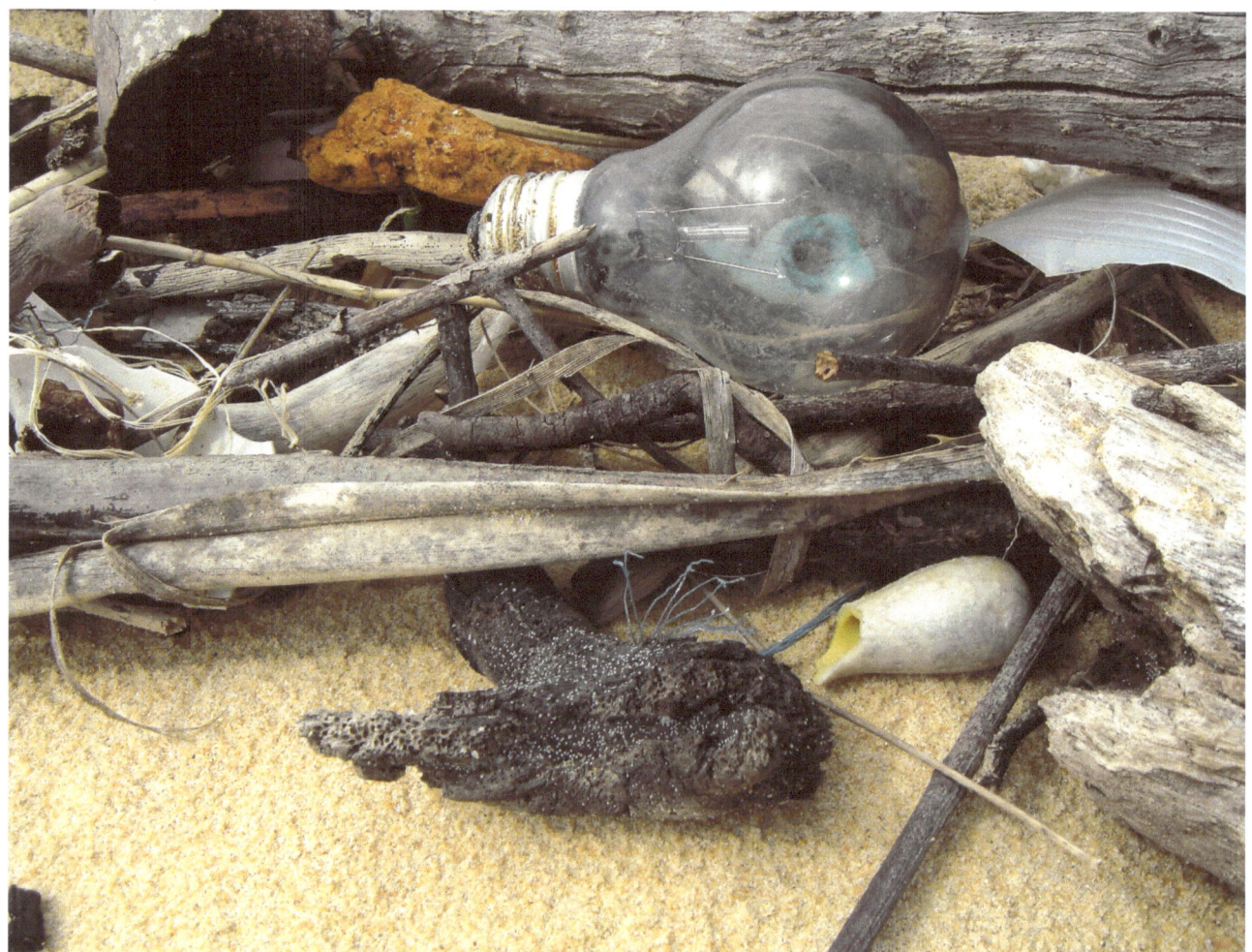

Disorder comes where learning stops

It was late night when we again passed Khao Lak's tourist centre. The opulent Scandinavian restaurant was still open, and still without customers. There were a few cars parked outside the nearby Italian restaurant and we could hear the volunteers young, beer influenced and loud, voices from its inside. I looked in the direction towards the sea where it was all dark, "No lights by the sea now. There will be little affluence left when also the young volunteers go back home." Chiraphorn too looked into the same dark by the invisible desolated shoreline, "No worries, the mess will soon be cleaned up and then life will come back again. After all, this is Paradise Beach."

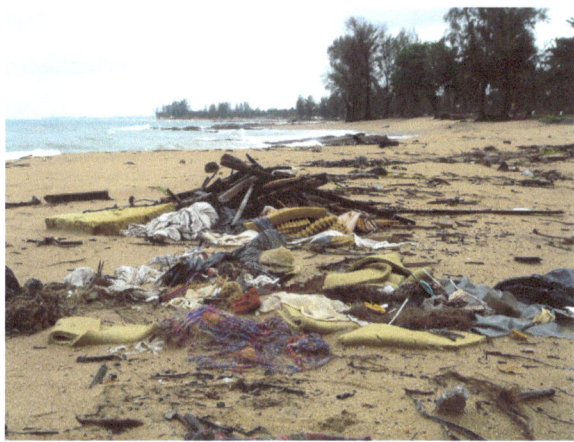

Piles of garbage a rainy day – gone by now

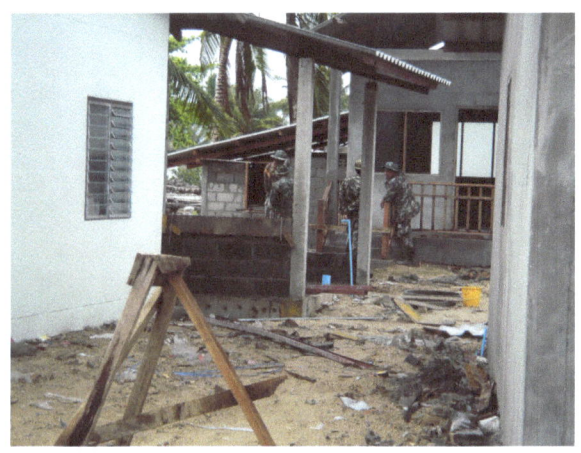

Military help to rebuild

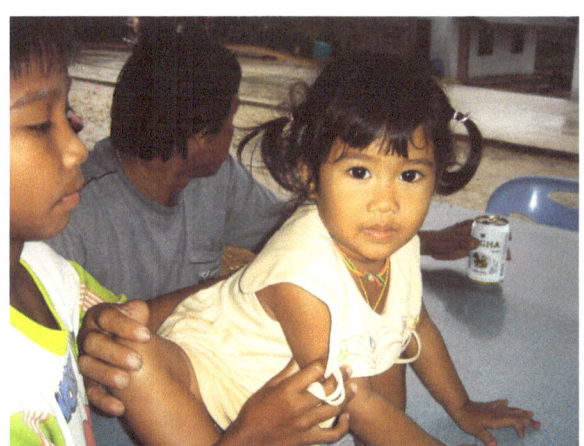

An adopted boy, one of many fathers, and a little princess

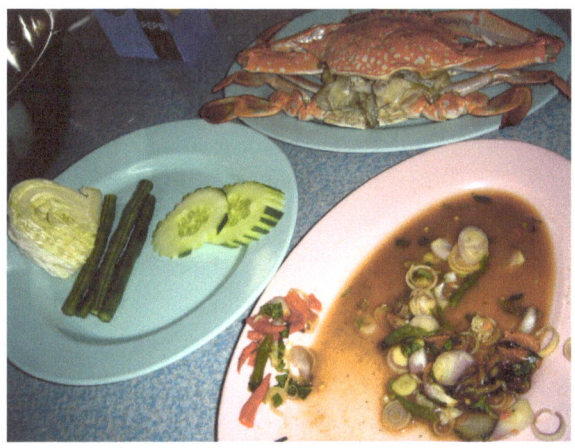

Eating in a village that once ceased to exist

www.ingramcontent.com/pod-product-compliance
Lightning Source LLC
Chambersburg PA
CBHW050353180526
45159CB00005B/2002